T0175229

Between
Philosophy
and Science

Between Philosophy and Science

Edited by
MICHAEL HELLER
BARTOSZ BROŻEK
ŁUKASZ KUREK

Copernicus Center
PRESS

© Copyright by Copernicus Center Press, 2013

Editing & Proofreading:
Aeddan Shaw
Piotr Godlewski

Cover design:
Mariusz Banachowicz

Layout:
Mirosław Krzyszkowski

Typesetting:
MELES-DESIGN

Publication supported by the John Templeton Foundation Grant
"The Limits of Scientific Explanation"

ISBN 978-83-7886-008-2

Kraków 2013

Copernicus Center
PRESS

Publisher: Copernicus Center Press Sp. z o.o.,
pl. Szczepański 8, 31-011 Kraków,
tel/fax (+48) 12 430 63 00
e-mail: marketing@ccpress.pl
www.ccpress.pl

Printing and binding by *Colonel SA*

Preface

The relationship between philosophy and science is intricately complex and notoriously difficult to describe. This is why the zone of interaction between them is so fascinating, both for philosophers interested in science and scientists with philosophically open minds. This "in between zone" can by no means be reduced to a narrow band separating these two fields; it rather infiltrates deeply the strata of scientific research and invades many areas of philosophical analysis. It is not only that science – as a certain phenomenon – may be an object of study for philosophers. By following some especially vital strands of thought, leading through the most fundamental scientific theories or models, one reaches domains on which one cannot avoid asking the kind of questions which are traditionally reserved for philosophy, questions such as those pertaining to existence, truth or rationality. However, less ambitious but equally important problems inhabit the "interface zone", for instance in problems related to space, time, causality and various epistemological questions. This zone is far from static; it is full of motion and truly dynamic. It is very instructive to investigate the migration of concepts between scientific and philosophical discourses and to see how these concepts change their meaning depending on their actual environment.

The volume we now offer to the Reader can be regarded as an excursion, or rather an exploratory expedition, into this "in between zone" and this constitutes a brief presentation of your companions in this risky enterprise.

The volume opens with a paper by Robert Audi, *Naturalism as a Philosophical and Scientific Framework: A Critical Perspective*, which tackles the problem of naturalism. After introducing some of the classical doctrines of naturalism, the author undertakes a critical analysis of scientific (methodological) naturalism, and attempts to substantiate the thesis that it does not necessarily exclude ontological pluralism. He argues further that the so-called scientific worldview does not require the assumption of the causal closure of the universe. Roman Murawski's *On Proof in Mathematics* is devoted to the elucidation of the concept of proof. The author claims that one should distinguish between two different ways of understanding 'proof' in mathematics: the 'informal' one, used in the every-day practice of mathematicians, and the 'formal' one, developed by logicians and those working on the foundations of mathematics. The following chapter, *Logical Form and Ontological Commitments* by Krzysztof Wójtowicz, poses the question of what is the relation between ontology and 'ideology' *vis a vis* the problem of the ontological commitments of scientific theories. The problem is discussed against the background of various interpretations of quantifiers: it turns out that one's chosen understanding of quantifiers influences one's ontological commitments. Bartosz Brożek, in *Neuroscience and Mathematics. From Inborn Skills to Cantor's Paradise*, provides an overview of the results obtained in the neuroscience and psychology of mathematics, and attempts to reconstruct the phylogenetic and ontogenetic history of mathematical cognition. He claims that mathematics is an 'embrained', embodied and embedded phenomenon. Against this background, he analyzes two classical problems connected to mathematics – mathematical Platonism and the unreasonable effectiveness of mathematics in the natural sciences.

Michael Heller's chapter, *The Ontology of the Planck Scale*, is devoted to the question of how the physical theories which aim at describing phenomena at the Planck scale may be used in ontological speculation. The author's claim is that since the Planck scale is currently the most fundamental level known to science, it should also teach us something about the fundamental ontological problems. In particular, at the Planck scale such notions as object, time, causality or change seem to be essentially different than the corresponding concepts in the classical ontologies of Aristotle, Kant or Whitehead. Wojciech Grygiel, in *Spacetime in the Perspective of the Theory of Quantum Gravity: Should It Stay or Should It Go?*, argues – against the background of an analysis of such research programs as the loop quantum gravity, superstring theory, twistor theory, topos theory or noncommutative geometry – that the categories of space and time, as we understand them, may play no role at the Planck scale, and hence it is most likely that those intuitive categories will disappear from the explanations of the fundamental phenomena. Helge Kragh, in *The Criteria of Science, Cosmology, and the Lessons of History*, considers what are the criteria which enable the distinction between science and pseudo-science. His analysis is carried out against the background of two examples: the steady state cosmology and the anthropic multiverse conceptions. He underscores the role of philosophical assumptions in the construction of both theories. The chapter culminates in the presentation of several examples of the uses and abuses of the history of science in arguing for or against particular scientific theories. The following chapter, Bogdan Dembiński's *Structuralism in Platonic Philosophy of Science*, aims at substantiating the thesis that, in the eyes of a philosopher of science, Plato may be seen as a structuralist. The author begins by identifying the key aspect of Plato's thinking, i.e. his account of the relationship between One and Many, as depicted in the dialogue *Parmenides*. Dembiński claims that for Plato each and every relationship between One and Many constitutes a whole, which in the contemporary terms may be called a structure. He then concludes with the description of the major features of Plato's structuralism.

The goal of Wojciech Załuski's chapter, *On the Relevance of Evolutionary Anthropology for Practical Philosophy*, is to assess the importance of evolutionary anthropology in the context of practical philosophy. The author argues that evolutionary anthropology –

understood as a view of human nature implied by the findings of evolutionary psychology – is more suited to constitute an important aspect of legal philosophy rather than of moral philosophy. The main argument backing this claim is that the primary object of moral evaluation are motives, while legal evaluation pertains to actions. Łukasz Kurek in *Emotions from a Neurophilosophical Perspective* addresses the question of whether emotions constitute a coherent category, both from philosophical and scientific perspectives. He begins by identifying various levels of the explanation of the concepts used in folk psychology (such as emotions), and indicates that the concept of emotion is both of philosophical and neuroscientific interest. Philosophy and science provide us with two popular theories of emotions, the cognitive and the noncognitive. The author claims that they both account for some phenomena usually categorized as emotions, which leads to the conclusion that emotions do not constitute a natural kind. Teresa Obolevitch, in her chapter *The Issue of Knowledge and Faith in the Russian Academic Milieu from the 19th to the 21st Century*, analyses the relationship between science and religion in the Russian ecclesiastical tradition. She underscores the peculiarities of the Russian academic philosophy, and describes the development of 'natural apologetic' in the 19th century, as well as the 20th century conceptions of the relationship between faith and reason.

The papers collected in this volume have been written within the research project *The Limits of Scientific Explanation*, carried out at the Copernicus Center for Interdisciplinary Studies in Kraków and sponsored by the John Templeton Foundation. It is our hope that these contributions – even if they take on different problems and make recourse to different philosophical methods – together constitute a coherent illustration of how complex and intriguing the relationship between philosophy and science may be.

Michael Heller
Bartosz Brożek
Łukasz Kurek

Table of Contents

Robert Audi
University of Notre Dame

Naturalism as a Philosophical and Scientific Framework
A Critical Perspective[1]

Introduction

In contemporary philosophy, naturalism is probably the most pervasive and influential intellectual orientation. It is widely presupposed; deviations from it are commonly felt to need justification; and where an important phenomenon, such as value, does not seem naturalizable, we find a plethora of attempts to show that, despite appearances, it is. Naturalism was important in philosophy even in the 19th century, but its popularity increased with the influence of certain elements in positivism and with the development of the philosophy of science. In 1969, W. V. Quine published "Epistemology Naturalized", which helped to popularize the verb 'to naturalize', and ever since then the philosophical literature has had title after title announcing a naturalization of something viewed as highly significant.[2]

[1] The publication of this paper was made possible through the support of a grant "The Limits of Scientific Explanation" from the John Templeton Foundation.

[2] *Epistemology Naturalized* appeared in Quine's collection, *Ontological Relativity and Other Essays* (Columbia University Press, New York 1969). For discussion of this and other versions of naturalism in philosophy, see *Philosophical Naturalism*, "Midwest Studies in Philosophy" 1994, vol. XIX, where we find *Autonomy Naturalized* by Marina A. L. Oshana, among other defenses of one or another naturalization project. The papers by Laurence BonJour, Richard Foley, Richard Fumerton, Richard Grandy, and Peter Hylton all have special relevance to appraising Quine's naturalism. See also David Papineau, *Philosophical Naturalism*, Blackwell, Oxford 1993. Recent treatments

But as common as appeals to naturalism are, its defense is rarely accompanied by a general account of the position, and it is frequently defended at best from a restricted point of view, such as that of epistemology or ethics. I am not prepared to offer a nicely detailed conception of naturalism as adequate to all of these contexts. But we can approach a plausible broad conception through reflection on how naturalization might proceed in some major philosophical territories: the metaphysical, including a focus on the mental; the epistemological; the scientific; the theological; and the ethical. I begin with some background observations; proceed to sketch an approach to conceptualizing naturalization projects and thereby naturalism as the successful completion of such a project. This will help us to see how much unity the notion of naturalism has, or can be given. A special concern of the paper is to determine what kind of naturalism, if any, is required as a presupposition of scientific inquiry.

1. The Historical and Philosophical Context

Historically, naturalism has ancient roots, perhaps most notably in Aristotle, who saw human beings as part of nature, derived an ethic mainly from what he took to be natural facts about us, and made no sharp distinction between scientific and philosophical knowledge. Aquinas's influential rendering of Aristotle preserved certain elements in Aristotle's naturalism, but denaturalized much of his thought, not only by making theology central and, I believe, introducing a notion of the a priori that went beyond the resources of Aristotle's apparent empiricism, but also by putting natural laws under God and licensing, though without mapping, a supernatural route to certain discoveries about nature. 17th century rationalism,

---

are found in M. De Caro, D. Macarthur (eds.), *Naturalism in Question*, Harvard University Press, Cambridge, Mass. 2004; S. Goetz, C. Taliaferro, *Naturalism*, William B. Eerdmans, Grand Rapids 2008; and D. Bradden-Mitchell, R. Nola (eds.), *Conceptual Analysis and Philosophical Naturalism*, MIT Press, Cambridge, Mass. 2009.

especially in Descartes, took philosophy still further from Aristotelian naturalism.[3]

Elements of rationalism remained in some of the British empiricists, at least Locke; but empiricism, especially in Hume, who carried it into semantics as well as epistemology, emerged as fertile ground both for the beginnings of modern naturalism and for the scientific worldview. Perhaps Kant may be seen as, in part, reconciling a kind of rationalism with a scientific worldview; but it was in England, particularly with J. S. Mill, and later, in America, that naturalism most prominently flourished in the nineteenth century.[4]

By the early 20th century, naturalism had strong proponents on both sides of the Atlantic and especially in America, where Roy Wood Sellars wrote, in 1922, "We are all naturalists now," though he quickly added – quite insightfully, in my view – that naturalism "is less a philosophical system than a recognition of the impressive implications of the physical and biological sciences."[5] Wilfrid Sellars, writing three decades later, went further. Parodying Protagoras, he said that "science is the measure of all things, of what is that it is and of what is not that it is not."[6] And in a famous paper defending materialism about the mental, J. J. C. Smart had no hesitation in saying "That everything should be explicable in terms of physics (...) except

[3] Spinoza is more difficult to locate in relation to naturalism than Descartes and Leibniz and might be argued to have, or at any rate make room for, a substantial naturalistic element.

[4] One should, however, note Comte and other French figures important in the history of positivism.

[5] Quoted by Hilary Kornblith in *Naturalism: Both Metaphysical and Epistemological*, „Midwest Studies in Philosophy", vol. XIX, no. 4, p. 50. Sellars was of course overlooking G. E. Moore and other non-naturalists in early analytic philosophy. Cf. Barry Stroud's comment, "'Naturalism' seems to me . . . rather like 'World Peace'. Almost everyone swears allegiance to it, and is willing to march under its banner. But disputes can still break out. . . And, like world peace, once you start specifying concretely exactly what it involves . . . it becomes increasingly difficult to reach and to sustain a consistent and exclusive 'naturalism'." See *The Charm of Naturalism*, [in:] M. De Caro, D. Macarthur, *op.cit.*, p. 22.

[6] W. Sellars, *Empiricism and the Philosophy of Mind*, [in:] *idem*, *Science, Perception and Reality*, Routledge and Kegan Paul, London 1963, p. 173 (originally published in "Minnesota Studies in the Philosophy of Science" 1956, vol. 1).

the occurrence of sensations seems to me frankly unbelievable."[7] He apparently intended 'everything' to apply unrestrictedly, but even excepting such abstract entities as numbers would make him a naturalist about what is commonly called *the world*.

20th century naturalism may have arisen initially as a reaction to supernaturalism, which posits beings or (concrete) entities that are neither in the physical world nor (assuming existence is possible for what is not in the physical world) necessarily physical or material at all. So conceived, naturalism stresses a non-theistic approach to understanding the world and, typically, scientific ways of explaining events. Modern empiricism is also a factor: its adherents see both meaning and knowledge as arising from our experience of the natural world in a way that makes it easy to conclude that the truths about nature are the only basic truths there are. 20th century positivism, though it refines this view, retains its naturalism: for positivism, observation and scientific method are the only grounds of substantive knowledge (knowledge of logical and analytic truths is not considered substantive on this approach).

Early in the 20th century, naturalism was often clarified by contrast with two major philosophical views. First, naturalists in philosophy opposed idealism—roughly, the theory that reality depends on, or is indeed constituted by, minds and their ideas. Second, and perhaps even more important, naturalists opposed ethical views that posited intrinsic value, such as that of the dignity of persons, as an irreducibly non-natural phenomenon. The rightness of an action, on these views—Moore's for instance—is a moral property that cannot be reduced to (e.g.) any naturalistic property, such as the total pleasure the action produces. On Moore's view, pleasure *has* intrinsic value, but is not equivalent to the intrinsically good.[8]

---

[7] J.J.C. Smart, *Sensations and Brian Processes*, "Philosophical Review", 1959, vol. LXVIII. The "frankly" here suggests a high level of conviction and also that he realized he was not able to provide a conclusive case for the mental-physical identify thesis.
[8] For a detailed examination of Moore's metaphysics of value see P. Butchvarov, *Skepticism in Ethics*, Indiana University Press, Bloomington 1989. Arguably, though a kind

More recently, naturalism has been contrasted with Cartesian dualism and interpreted in relation to the possibility of an ultimately physicalistic understanding of human behavior. We are flesh and blood, neurons and synapses; and if the life of the mind is more than a powerless shadow, it must occur in this biological network. In the past three decades, naturalism has also become a powerful force in the theory of knowledge and justification. Knowers are seen not as meeting evaluative standards, such as criteria that *entitle* one to believe, but as reliable instruments responding to experience of the world: they reliably register truth rather as a good thermometer registers temperature.

In metaphysics, then, we have naturalism both as a reaction to supernaturalism and as a rejection of mind-body dualism such as we find in Descartes; in ethics, naturalism—at least in realist, as opposed to noncognitivist versions—appears as a rejection of irreducible notions of value and rightness (and an effort to naturalize the corresponding properties); and in epistemology naturalism accounts for knowledge and justification in terms of notions amenable to scientific treatment, particularly the concepts of physics and psychology. If there is a unifying conception of naturalism in all these domains, it is at least not commonly made clear. One could, for example, be a naturalist in rejecting transcendent beings, but a non-naturalist in the philosophy of mind, ethics, or epistemology. Indeed, it may be that, in any one of these domains, there are forms of naturalism that are consistent with rejecting naturalism in all the others. If, however, we recall Sellars's claim that science is the measure of all things and Smart's view that everything is explicable in terms of physics, we may think of philosophical naturalism as, in outline, the view that nature is all there is and the only basic truths are truths of nature. This is certainly not the only kind of naturalism that merits the term 'philosophical', but it

---

of physicalism often associated with scientific realism is a pervasive element in twentieth-century naturalism and at least equally at present, anti-realism in metaphysics, unlike the thesis of the irreducibility of the normative, is compatible with naturalism.

represents a strong, broad thesis that serves as a good focus for analysis and comparison.[9] We can gain a better understanding of naturalism so conceived—that is, as a broad metaphysical thesis—and of departures from it by considering some of the projects philosophers regard as advancing it.

2. Scientific Naturalism

Many philosophers of science today hold or presuppose some form of philosophical naturalism. But are scientists as such committed to any form of it? Here we should scrupulously observe the distinction between philosophical naturalism and methodological naturalism. *Methodological naturalism*—which, unlike the former, is not a metaphysical position—is, roughly, the view that causes and explanations of natural phenomena should be sought in the natural world, paradigmatically in terms of what meets at least three commonly endorsed criteria for a scientific theory—testability, publicity, and empiricality (these are taken to be important necessary conditions, not sufficient conditions, though a case can be made that there is an interpretation of them on which anything satisfying all three and deserving the name 'theory' would be scientific). Methodological naturalism requires conducting scientific inquiry in what might be called descriptive natural categories and with testability and the experimental, public, predictive character of science in mind; but it does not imply a comprehensive ontology nor even deny the existence of anything supernatural. Indeed, methodological naturalism is not fundamentally descriptive at all; it is prescriptive, a normative (though non--moral) claim about scientific inquiry. It is clearly consistent with Roy Wood Sellars's contention that naturalism is "a recognition of the im-

---

[9] For a broader view that constitutes a philosophical naturalism, see the "liberal naturalism" set forth by M. De Caro in *Naturalism in Question* and in M. De Caro, A. Voltolini, *Is Liberal Naturalism Possible*, [in:] *Naturalism and Normativity*, eds. M. De Caro, D. Macarthur, Columbia University Press, New York 2010, pp. 69–86.

pressive implications of the physical and biological sciences." This contention does not imply full-scale philosophical naturalism, even if for some thinkers, especially those who are pragmatically inclined, they make it attractive.

Naturalism and the drive for intellectual economy

It must be granted that naturalism as I have characterized it (and also as it is usually described) is no clearer than the notion of the natural, but a standard assumption is that the physical is paradigmatically natural and, to take the strongest contrast, that *God* as understood in any major monotheistic religion is supernatural. The elements of scientific method I am emphasizing are methodological, not ontological. As such, they are neutral with respect to the possibility of supernatural agency, and methodological naturalism neither affirms nor denies theism. Neither the testability nor the publicity nor the empirical character of scientific hypotheses entails any substantive ontological conclusion, such as the thesis that only natural events can be causes or that there cannot be an infinite chain of causes extending into the past.

It is, however, commonly and plausibly held that a commitment to scientific method presupposes a commitment to some form of Ockham's Razor, understood as a principle calling for intellectual economy. Does this principle favor philosophical naturalism—or at least the anti-supernatural elements in it—over methodological naturalism, which is neutral with respect to theism? The question is not easy to answer. On analysis, the status of Ockham's Razor is not altogether clear. On any interpretation, it implies that other things equal, the simpler of two competing views is preferable. Suppose this is our working formulation. Isn't philosophical naturalism a simpler worldview than, say, a theistic one that is otherwise as much as possible like, say, the scientific realism that a philosophical naturalist could countenance? Answering this requires some account of simplicity, an

elusive notion I cannot take time to explicate.[10] Still, if we take it
that the number of basic assumptions and irreducible kinds of entities
posited by a worldview is most important for determining the degree
of its simplicity, then we might suppose for the sake of argument that
there is *some* reason to think that the naturalistic worldview is simpler than a theistic view.

Suppose the naturalistic worldview is simpler than any plausible
theism. Some philosophers would stress that it still provides no answer to the classical cosmological question, Why is there something
and not nothing? If it does not, and theism does, then other things
are not equal in the comparison between, on the one hand, a theistic worldview combined with methodological naturalism and, on the
other side, philosophical naturalism. And, for many lives, philosophical naturalism does not accommodate the most plausible explanations for certain religious experiences or, indeed, for some people's
long-standing patterns of religious experience. Either of these points
may well suffice to outweigh even differences in simplicity *greater*
than those that *may* exist between a classical theistic worldview and
naturalism.

In the light of these points, I doubt that Ockham's Razor—understood as the principle that in framing and adopting explanations
and theories, we should prefer the simpler, other things equal—cuts
deeply against a classical theistic worldview as such. Note that the
principle does not say, and in my view should not be taken as entail-
ing, that it is *irrational* to hold a view while aware that an alternative that is otherwise as plausible is simpler. Rational preferability
of one view over another does not imply the irrationality of holding
the (perhaps marginally) less desirable view. As to preferability, we
should consider at least two kinds that bear on comparing hypotheses and worldviews.

---

[10] Richard Swinburne argues in detail that the theistic view is simpler. For an indication
of his view of simplicity, see *The Existence of God*, Oxford University Press, Oxford
2004, pp. 53–66, 70–71, and 179–179; and *Epistemic Justification*, Oxford University
Press, Oxford 2001.

Two conceptions of the Razor as endorsing preference
for the simpler

So far, I have spoken as if it were clear what the status of the Razor is.[11] But it is not. Once this is seen, the bearing of the principle on the preference for a naturalistic worldview can be clarified. The principle is most widely considered epistemic, in the broad sense that it provides an evidential standard relevant to assessing the candidacy of a belief, hypothesis, or theory for justified belief or for knowledge or for both. It also applies to *acceptance* as a propositional attitude or to hypotheses as candidates to be objects of any kind of epistemic attitude. To assess this principle, we should take account of two major approaches in epistemology, internalism and externalism. Accordingly, we must distinguish internal from external epistemic considerations: roughly, between justificatory elements—evidences, in a sense—that are accessible internally, i.e., to reflection or consciousness, and, on the other hand, objectively truth-conducive factors that are not thus accessible.[12]

May we (other things equal) take the simpler view to be better justified, on internal grounds? This may seem obvious. It is not. What may be obvious is something easily conflated with the principle conceived epistemically: namely, the quite different principle that, other things equal, it is preferable to *work* with a simpler view in constructing a theory or in framing or retaining a worldview. There is less to deal with—to explain, to connect with other elements, and so forth. That interpretation, however, yields a practical principle that could be true even if there were no difference in evidential status between

---

[11] For a brief discussion of how to interpret the razor principle and its possible role in evaluating design arguments, see R. Collins, *The Design Argument: Between Science and Metaphysics*, [in:] *Analytic Philosophy Without Naturalism*, eds. A. Corradini, S. Galvan, E.J. Lowe, Routledge, London 2006, pp. 140–152.

[12] For a concise but wide-ranging discussion of the internalist-externalist controversy, see my *Epistemology: A Contemporary Introduction*, London and New York, Routledge 2010, esp. chapters 10–11. These chapters also indicate why the notion of knowledge may be plausibly represented along reliabilist lines that do not require taking it to entail justification. If justification is not naturalizable, then, it does not follow that knowledge is not.

more and less simple views. Call the practical principle in question *the principle of least effort*. It might also be called *the principle of optimal efficiency* to emphasize that it opposes expending more effort than is needed for one's governing purpose.

This principle of least effort may be self-evident, and it is plausible from an internalist point of view (at least partly on the a priori basis of understanding alone); but it does not entail the Razor principle as stated epistemically and understood internalistically. A hypothesis that is easier to work with is not thereby better evidenced, say in terms of confirmatory experiences or predictive power, than an otherwise equally well confirmed alternative.

A return to the Razor principle is considered to be epistemic: roughly, the principle of the greater likelihood of the simpler hypothesis. We have seen no reason to endorse it from an internalist point of view. Do we have reason to think that from an *external* perspective, simpler views are more likely to be true? Perhaps, for some plausible notion of simplicity, we do, given common experience and the track record of scientific hypotheses. The intuitively simpler hypotheses have apparently tended to be better confirmed (though this might be at best hard to show owing to the difficulty of separating considerations of simplicity from those of degree of confirmation). I do not see how to demonstrate this, though doubtless we might entertain an evolutionary explanation. The preference for simplicity could have fitness value: perhaps, since we do in fact generally prefer simpler hypotheses, we probably would not have survived, or at least survived and reached the level of efficiency we have achieved, if the ways of nature were not better evidenced by simpler hypotheses.

It is important to see why this evolutionary hypothesis provides at best limited support. First, it does not entail the truth of the Razor principle understood epistemically. Second, it presupposes that we have perceptual knowledge or at least true perceptual beliefs in the first place, in which case the question whether simplicity enters into *those* notions would have to be addressed before that justification could be considered to be sufficiently independent of simplicity to support the Razor without a vitiating circularity. Such inductive reasoning in favor of the Razor taken epistemically also provides only an empirical justification. I propose simply to assume that nature is (other things equal) better described by simpler hypotheses and thus to grant that the Razor may have at least *external* evidential force. What might explain why it does—why nature's ways are better evidenced, in terms of objective probability, by simpler hypotheses?

Another possibility is that nature might be *designed* to be understood and predicted by beings like us who prefer simpler hypotheses.[13] I see no way to show that this (presumptively theistic) explanation is correct, but naturalists should not simply rule it out of hand, particularly since there need be no conflict between a theistic worldview and a deeply scientific habit of mind. The world can be studied with as much scientific rigor by those who believe it is divinely created as by those who think it is a fortuitous precipitate of blind forces.

3. Ontological Pluralism

So far, we have seen no decisive reason to think that a scientific habit of mind requires commitment to philosophical naturalism, as opposed to an ontologically more pluralistic worldview. For some highly scientific people, however, including those who are theists, methodological naturalism is a possible position—even a comfortable position. We have also seen no decisive reason to consider theism clearly less plausible than atheism given a commitment to Ockham's Razor. Even if it is less simple in some way, it does not follow that it is less well evidenced. Ontological pluralism, at least of a kind affirming the reality of both natural and non-natural entities, seems a plausible position. To be sure, it should be emphasized that even a kind of materialist

---

[13] Cf. Robert Koons on *The Incompatibility of Materialism and Scientific Realism*, in his *Realism Regained*, Oxford University Press, Oxford 2000, chapter 17, section 5.

ontology is compatible with a theism on which God is embodied in matter—as possibility that also seems open, however unorthodox it may be.[14]

Abstract entities

A full-scale materialist ontology must also discountenance abstract entities. This is a prospect I cannot pursue in detail, but philosophical naturalists are in any case more concerned with naturalizing any reality with causal power (as supernatural beings are thought to have) than with naturalizing the abstract, which is typically considered to be outside the causal order. Abstract entities are a kind of annoyance to some naturalistic thinkers, since they must be accounted for in a comprehensive ontology; but if they have no causal power, they at least do not *do* anything in the natural world and cannot compete with scientific explanations or undermine scientific predictions.

Monistic naturalism—the kind countenancing only one basic ontological category, such as the physical—seems to me quite implausible even as applied only to bearers of truth value. If, for instance, they are propositions, or even sentence types as opposed to sentence tokens, then there are some abstract entities.[15] Sentence tokens can be argued to be physical—at least if we can plausibly take mental tokens of sentences to be physical—but sentence types of the kind represented by what is common to two printings of the same article—are not plausibly viewed as physical. This is not to suggest that we should not avoid pos-

---

[14] This possibility is described in some detail in ch 10 of my *Rationality and Religious Commitment*, Clarendon Press, Oxford 2012.

[15] For an excellent but too little studied paper on this issue see R. Cartwright, *Propositions*, [in:] *Analytical Philosophy*, ed. R.J. Butler, Blackwell, Oxford 1962, pp. 81–103. Among its points is the important one that sentences and propositions (or any bearer of truth value) differ in their *arithmetic*. The number of things said is determined differently than the number of sentences in which it may be said. If two people who, respectively, speak only English and Polish, say in their own languages that $7 + 5 = 12$, there is one thing said and one truth, but there are two sentences.

iting abstract entities, other things equal—as Ockham's Razor on any plausible interpretation would counsel. If sentence types could do the work of propositions and the former are unavoidable posits, then we could treat talk of propositions as a mere convenience. But sentence--types are also abstract. Still, neither posit necessarily affects scientific inquiry; if scientists do not find proposition talk burdensome— or prefer it to sentence talk as less likely to result in ambiguity—philosophers should feel no need to discountenance propositions on the ground that positing them is at odds with good scientific practice.

Universals and numbers present other challenges, but there, too, I must be content simply to point out the obstacle to full-scale materialism rather than argue for commitment to pluralism that includes the abstract. Suppose, however, that there are apparently properties that are not even descriptive—and in that minimal sense natural—to begin with. Then there would seem to be a twofold problem for a thoroughgoing naturalism: first, to eliminate the normative element; secondly, to eliminate the abstract element. Here, I will concentrate only on the first problem and then only in the general way appropriate to a broad essay of this kind.

The normative domain

Supposing there are normative properties—which, to be sure, non-cognitivists deny—can reductive naturalists accommodate them? The Cornell Realists have attempted to give a kind of reduction here, but I have elsewhere argued that although we might regard them as providing reasons to think *moral explanations* can be naturalized, they have not shown that moral properties are naturalizable.[16] Briefly, the idea is that when a moral property, such as the injustice of a regime,

---

[16] I appraise one version of Cornell Realism, and sketch some conceptions of moral explanation in *Ethical Naturalism and the Explanatory Power of Moral Concepts*, [in:] *Naturalism: A Critical Appraisal*, eds. S. Wagner, R. Warner, University of Notre Dame Press, Notre Dame 1993, pp. 95–115.

is cited to explain behavior, say a revolt against that regime, the explanatory work is done by some grounding property that *is* natural. For instance, wherever injustice can be justifiedly and correctly cited as explaining a revolt, the person giving the explanation must be justified and correct in taking to be present such grounds of injustice as police brutality and virtually confiscatory taxation (say approaching ninety percent), and *these* are natural phenomena that can explain a revolt in terms of social-scientific generalities that do not require normative concepts in their content. Moreover, if we did not take injustice to imply the presence of such plainly causative elements, we would not see how it explains the revolt. It does not brutely explain that, even in the way releasing a soccer ball in midair might brutely explain its falling. We can detail, and must be able to detail at least by specifying grounding elements, *how* normative ascriptions succeed in explaining what they do explain, in a way we need not be able to do this for the dropping ball or, in the psychological case, for the normal raising of a hand, given the agent's decision to raise it.

There is, however, another route to possible naturalization of the normative. Suppose for the sake of argument that we can formulate a disjunction of all the grounds of moral obligation. If W. D. Ross's famous list is as comprehensive as he hoped, we could say that an act is obligatory (for a person) if and only if (in rough terms) it is *either* a doing of justice (in the descriptive sense of treating persons equally), an avoidance of injury, a promise-keeping, an avoidance of lying, a beneficent deed, an act of self-improvement, a reparation for injury, or an expression of gratitude. One could do the same kind of thing in epistemology, for instance holding that a belief is justified if and only if is grounded in either perception, or memory, or introspection, or reason, or testimony. Even if these biconditionals are not analytic or any kind of conceptual truth,[17] might they suffice for reduction of the normative properties in question to descriptive properties?

---

[17] As suggested in the text, if we had a *complete* list of the things that constitute a person's obligation, we might say that to be obligated to A is equivalent to satisfying one

Property identity as an element in ontological reduction

To answer this, we must at least assume that the equivalence is necessary. But may we take necessarily equivalent predicates to express the same properties—identical properties? This is doubtful. Consider the property (possessed by every other integer) of being even. Is this the same property as that of being a member of the series 2, 4, 6, 8, etc.? The property of being even (being an even integer) is *equivalent* to the property of being divisible by 2 without remainder; but is that property identical with, or even equivalent to, the property of being a member of the set 2, 4, 6...? The *members* of the set in question are thus divisible; but is being a member of it the same property as being divisible by 2 without remainder? The former is a set-theoretical property, the latter a numerical (arithmetical) one. These seem different in kind.

This example and others, such as those involving purported reductions of normative concepts or properties, bring out an important point. The relevant sense of 'reduction' is no clearer than our understanding of when two predicates express the same property (or, given any properties F and G, under what conditions $F = G$). This remains a vexed question. Do two predicates express the same property provided they are (a) synonymous, (b) analytically equivalent, (c) conceptually equivalent (if that is not equivalent to one of the other categories), (d) logically equivalent, (e) metaphysically equivalent, (f) synthetically a priori equivalent, (g) explanationally equivalent, (h) nomically equivalent, (i) causally equivalent, or (j) something else again? Even if we have an answer to this question, the ten notions in question remain controversial, and they all need analysis.

Whatever our criteria for property identity, it is far from clear that there are disjunctive properties *at all*, as opposed to disjunctions of

---

or more of the descriptions. Depending on the character of the list, we might consider this a (disjunctive) analysis. But one might also insist on some unifying element, such as the notion of respect for persons, and go on to argue that here such an element must be normative. More is said below about the possibility of a disjunctive analysis.

predicates that each express a property. If there are disjunctive properties, we need an explanation of why specifying that a shape is either circular or elliptical is not an answer to the question 'What are its
shape properties?' or, especially, 'What is its shape?' One could argue
that the needed explanation is simply pragmatic, say that in asking for
a shape we presuppose in normal contexts that a non-disjunctive specification will be given; but that is not confirmed by our explanatory
or descriptive practices even when answering a particular question in
a communicative situation is not a requirement for appropriateness.[18]

   To be sure, the disjunctively specified grounds of obligation cited
above are significant; they yield multiple paths from certain descriptive notions to the crucial notion of obligatory action. Similarly for
the grounds of justification, which might be argued to be disjunctively
specifiable in terms of, say, the classically recognized basic sources
of justification: perception, memory, introspective consciousness, and
intuitive reason. But this point about the grounding of the normative shows—assuming the grounds in question are genuinely constitutive—only that the normative properties (and the concepts of them)
are anchored in the natural world, not that those properties are genuinely natural or even equivalent to disjunctive properties, if there are
such properties.

   This is a good place to note that what has been said of properties
may well not apply to concepts. Concepts are essentially connected
with subsuming things under the abstract equivalent of a description,
an equivalent that may be disjunctive or otherwise compound. This
does not require indicating something about what the relevant things
are "like", especially where this is related to explanation, prediction,
connections to other things, or all of these (and perhaps other ele-

---

[18] To be sure, there is a sense in which for any true proposition, there is a possible
question to which it is a correct answer. My point here is that if the question is what
properties a thing has or what properties of it explain something, it is at best atypical for
a plausible answer to cite something that seems a genuinely disjunctive property. The
issue is a large one, however, and what is said here is meant only to suggest problems
that prima facie favor denying that predicative disjunctions of the kind in question express disjunctive properties.

ments). Viewed in this way, at least, concepts may be seen as disjunctive. In this, and in the very fine-grained way in which we individuate them, concepts are more like propositions than like properties.

One might now wonder whether we might naturalize normative concepts like that of obligation disjunctively, even if not the property. First, it is doubtful that a naturalistic concept can express a non-naturalistic property. If this is so, then conceptual naturalization of the normative will apparently fail if naturalization of normative properties fails. But even apart from this question, the concept of obligation is surely not equivalent to the disjunctive concept of being either a promise-keeping or an avoidance of harm or… for any purportedly complete list of the constitutive sources of obligation. These elements may perhaps indicate all of the various grounds of obligation but not what, conceptually, obligation *is*. It is a kind of commitment, and the disjunction indicates ways of incurring it, but not what it is.

4. Non-Reductive Naturalism

The view that non-natural properties are strongly supervenient on natural ones is sometimes called non-reductive naturalism, and the name is doubly appropriate if the supervenience is paired with a grounding relation. It normally is paired with one, though in the mental-physical case the grounding relation is empirical and presumably nomic, whereas in the normative case it is (on my view and that of many others) a priori and metaphysically necessary rather than nomically necessary. Such naturalism is, however, clearly a liberal naturalism: although it may embody a denial of substance dualism—a view that liberal naturalists reject—it is consistent with property dualism, since even the strong grounding relation obtaining in the normative cases does not imply the possibility of reduction of the normative to the natural. The point is not that grounding is not a reductive relation, though that is true and important; the point is that this character of the relation does not imply the impossibility of reduction of the grounded properties to *some* property of the kind to which the grounding property belongs, in this case natural properties.

It should be added that so-called non-reductive materialism as so far described is even consistent with Cartesian dualism. If mental properties are grounded in (but not reducible to) physical ones—the kind of natural property important for naturalism in the philosophy of mind—then the mental properties can be in a different category, perhaps being properties of a Cartesian mental substance. To be sure, mental-physical causal relations may be less easy to understand given that picture, but nothing in the concept of causation precludes cross--categorial causal relations. I would stress, however, that people who hold non-reductive naturalism would generally restrict mental properties to properties of embodied beings and would not countenance Cartesian mental substances. One metaphysical problem, of course, is how to understand mental properties as at once not reducible to physical ones yet *necessarily* belonging to physically embodied beings.

Even if mental properties are not themselves physical, psychological science can proceed as it would if they were physical. If, for instance, there are lawlike connections between, on the one hand, sets of desires and beliefs and, on the other hand, actions, these can be studied either (a) psychologically in relation to objective criteria for applying the relevant psychological and behavioral predicates or (b) neurophysiologically in terms of the grounding properties and their role in producing either the psychological states or the physical movements those states explain, or both. For the normative counterpart of non-reductive physicalism, the point is different, since some of the generalizations in question, say that failure of the prosperous to do any beneficent deeds is wrong, are not empirical. But some of the generalizations are empirical—say that injustice breeds anger—even if the empirical work, and hence the task for social-scientific appraisal, is at the level of the grounding properties. If ethics is not itself a science, there is certainly a host of scientific questions about the conditions for its realization in human conduct, the risks to its grip on human motivation, and the identification of those for whom it yields guiding standards—moral individuals—and those for whom it does not.

It is the metaphysician, not the scientist, who tends to feel a need to reduce non-natural properties to natural ones. There is no reason to object to the quest and much to be learned from appraising its degree of success. But, as a philosophical framework appropriate to a scientific habit of mind and supportive of the progress of science, methodological naturalism is as strong a naturalistic position as required.

5. Realist and Anti-Realist Naturalisms

I have already suggested that eliminatism regarding abstract entities is implausible and that there is no need for it on the part of scientific investigators. But I have also suggested that normative properties do not have causal power, even if their presence can explain certain facts, as where people's lying and cheating explains why they are immoral. Let us explore the significance of these points for a representative kind of normative property, the moral.

Moral epiphenomenalism and ethical noncognitivism

Suppose that moral properties are not causal. This does not imply that they are epiphenomenal. Their instances are not *effects* of instances of the properties that ground them: instead, the former instances bear a consequential relation to those grounding instances. Being an instance of a lie does not cause being an instance of prima facie wrongness; it conceptually entails it. If wrongness is not an effect (strictly, an effect-property), it need not be epiphenomenal: indeed, on my view it is not even a candidate to be so. Epiphenomena are elements that figure in the causal order as effects (or effect properties) but abruptly—and, some would say, inexplicably—terminate a causal sequence.

The most plausible conclusion to draw here is probably that moral properties and concepts do not function, or do not mainly function, in the causal order, at least as causal elements therein. Nonetheless, it is in part because some philosophers believe that moral properties do not have causal power that they want to eliminate them.[19] Perhaps many naturalistic philosophers believe or presuppose that to be *is* to be the value of a causal variable. If moral properties lack causal power, can they be eliminated?

This is an indirect way of raising the question whether noncognitivism in ethics and other normative disciplines is successful. I do not see that it succeeds, but there remains a great deal of controversy about this. Perhaps the best argument for the view, an argument provided originally by Hume in the *Treatise*, proceeds from the premise that moral judgments, but not cognitive ("descriptive") ones are intrinsically motivating. I can agree that non-normative (roughly, descriptive) judgments are not intrinsically motivating, but I have argued (as have many others) that moral judgments are not either.[20] Rather than pursue this complicated issue, I prefer to indicate how my own ethical theory can accommodate elements of naturalism.

Moral realism in the intuitionist tradition

My strategy in metaethics—now called "the New Intuitionism",[21]—has not been negative even in containing a critique of noncognitive analyses. Instead, I have proposed a moderately rationalist intuitionism which anchors moral properties in natural (descriptive) ones, as

---

[19] To be sure, anti-realists in ethics would say that since there are no moral properties, it is the predicates used to express them that should be said to be non-causal, but we need not dwell on this.

[20] In my *Moral Judgments and Reasons for Action*, in my *Moral Knowledge and Ethical Character*, Oxford University Press, Oxford 1997, which contains many references to relevant literature.

[21] See J.G. Hernandez (ed.), *The New Intuitionism*, Continuum, London 2011, for papers articulating and critically appraising this view.

with the list of grounding properties from Ross (to which I have added two others[22]). This is a realist view, but does not require reducing normative properties to descriptive ones. Here I want simply to indicate that even if one could show that moral properties are natural, the major elements in my theory would be sustainable: the theory accords well with, but does not depend on, a non-naturalist account of normative properties. As is well known, G. E. Moore held that normative properties are non-natural. My intuitionism does not presuppose this. The most important point here is that the epistemic status I attribute to intuitive prima facie duty principles—being necessary and a priori—does not require that the properties figuring in those principles be non-natural.[23] Nor is that incompatible with or entailed by the thesis that moral properties are, as Moore, Ross, and other intuitionists also hold, consequential on natural ones. This neutrality toward Moorean non-naturalism is an advantage in making a version of intuitionism eligible for endorsement by many contemporary moderate naturalists, certainly including liberal naturalists.

If it turns out, then, that moral and other normative properties are natural, my view does not require either changing my position on what properties ground them—these are natural properties on the present account—or moving to anti-realism, or altering the relation I have described above between ethics and science. Nor need the moral epistemology be revised. The relation between moral and non--moral properties and concepts can have the character I have maintained whether or not those properties are natural. The metaphysical change from neutrality toward naturalism to endorsing some version of it would leave both the normative content and the epistemology of the view essentially unchanged.

---

[22] See ch 5 of my *The Good in the Right: A Theory of Intuition and Intrinsic Value*, Princeton University Press, Princeton 2004, for a description of the added grounding properties as well as refinements and corrections of Ross's list of such properties.

[23] There is even a way for my view to be adapted to noncognitivism, as I noted in *The Good in the Right*, p. 151. But noncognitivism is not, to be sure, a natural option for an intuitionist.

Noncognitivism is only one of many anti-realist positions. Not all of these are motivated by philosophical naturalism. Eliminativism in the philosophy of mind might be, since it discountenances mental properties in favor of mental predicates somewhat as occurs in ethical noncognitivism with moral properties and moral predicates, and it seems motivated largely by the sense that the physical is what is fundamentally real. But with eliminativism in the philosophy of mind we have an error theory, on which it is simply a mistake to hold that there are, for instance, beliefs. By contrast, ethical noncognitivism does not take moral claims to be truth-valued, hence does not view them as invariably erroneous. Scientific anti-realism is not motivated by taking theoretical entities to be, like irreducibly normative properties, inappropriate to be countenanced by naturalism or, so far as I can see, even by a physicalist interpretation of reality. The position is apparently motivated by Ockham's Razor and considerations internal to the task of understanding scientific explanation. In any case, in concluding this essay, I propose to consider a larger issue concerning the overall conception of the universe required by science as compared with the conception favored by philosophical naturalism.

6. Scientific Naturalism and Causal Closure

For thoroughgoing philosophical naturalists, although it is not an a priori truth that mental properties depend on physical ones, the physical has a kind of metaphysical sovereignty. One manifestation of this sovereignty concerns causal power—an element that, on any plausible metaphysics, entails ontic reality on the part of anything that possesses it. Let us begin with that.

Causal closure and the explanatory exclusiveness of the physical

For naturalism the causal power of mental properties must depend on that of some physical property at their base: even overdetermination by the joint power of physical and mental phenomena seems ruled out, or at least appears to be beyond the pale for a robust naturalism, which seems committed to a causal closure thesis on which all causes are physical, where this is understood to rule out that mental causes, unless *identical* with physical ones, can even be overdetermining causes of any effect. The view leaves open that *effects* of physical causes may be non-physical, since, if they are epiphenomenal, they might be merely ontologically an annoying dangler, but otherwise innocuous. Still, if a mental property has causal power, then on the closure thesis just stated, it is reducible to a physical property.[24]

This closure thesis may seem to be supported by the idea that physically identical worlds must be mentally identical. For it may appear that the only explanation of this plausible idea is either that mental properties *are* physical or the mental as such has no causal power. But that is not so; non-reductive materialism, especially if it embodies the idea that mental properties are grounded in physical ones, suffices to explain the idea. Consistently with that, however, is the not implausible view that every mental event has a description under which it is physical—a view that leaves open the relation between mental and physical *properties*. Taking the two together, some philosophical naturalists might hold that a mental event "as such" has no causal power.[25] Must a scientific naturalist hold this, however? It

---

[24] As Jaegwon Kim has put it, "One way of stating the principle of physical closure is this: If you pick out any event and trace out its causal ancestry or posterity, that will never take you outside the physical domain. *That is*, no causal chain will ever cross the boundary between the physical and the nonphysical." See *Mind in a Physical World*, MIT Press, Cambridge, Mass. 1998, p. 41 (italics mine). For extensive critical discussion of this and other closure theses (particularly as represented in this and later work by Kim), see S. Goetz, C. Taliaferro, *op.cit.*, esp. chapters 2 and 3.

[25] Elsewhere I have exposed a triple ambiguity, showing that none of the three readings depends on *identifying* mental properties with physical ones. See, e.g., *The Theory of Explanation: Some Metaphysical, Scientific, and Intentionalistic Dimensions*, forthcoming.

is one thing to require a physical ground as a platform for the causal power of the mental; it is quite another to maintain that only the physical has causal power.

We should ask, then, how important it is for a scientific naturalism to embrace the causal closure of the physical world, as opposed to its *causal sufficiency,* which constitutes a kind of *causal plenitude*: roughly, the existence of a physical nomically sufficient condition for every physical event. Causal sufficiency would assure the possibility of a comprehensive physical science. All physical events (including mental events, at least if mental properties are themselves physical) would be physicalistically explainable: roughly, explainable by citing (under an appropriate description) the instantiation of a physical property and in that sense explainable in physical "terms". On a stronger causal sufficiency view, all *natural* events—which would include any mental events in the natural world—would be physicalistically explainable. But even adding this clause would not yield an implication that mental events have no causal power. Let us explore the causal sufficiency view further.

Causal sufficiency and the unity of science

Suppose that every physical event is explainable by appeal to a law of nature. We still cannot assume that there must be a way to *unify* our laws—a possibility that, for most naturalists, represents an ideal—or even a way to produce intuitively appealing or technically useful theories. Ontological constitution of the world is one thing; its pattern of laws is another—nor does its even being governed by laws clearly follow from the nature of its constitution alone. Even if we assume that there are no naturalistically inexplicable physical events, causal closure by itself does not guarantee those scientifically desirable results—especially overall unification of laws. Causal sufficiency, by contrast, even in the strong form, would not be exclusive in the way causal closure is.[26] It would not, for instance, rule out mental *properties* as such having a measure of causal and explanatory power, in a sense implying that an event's having a mental property is sufficient for citing that event (under the relevant mentalistic description) to explain certain other events, such as actions.

To see a major difference between causal sufficiency in either form and causal closure, consider their relation to theism. Both forms of the sufficiency view are consistent with the possibility of *one* kind of supernatural causation. I do not mean the possibility that, at some point in time, God (or some god) created the universe with its laws and initial events, so that divine creation—which may be viewed as a kind of *originative causation*—in a sense *underlies* all causation, and each cause is a link in a causal chain tracing to creation at its origin.[27] For in that case the initial event in each causal chain would not have a natural cause. But I am including both the possibility of divine conserving causation of the universe from eternity (in which case there is no originative causation) and also the possibility of overdetermination in which a supernatural event, such as a divine volition, is sufficient, as is some natural event, such as a neural one, to produce another natural event, such as a sudden recovery of eyesight.

This is not to suggest that, on the sufficiency view conjoined with the view that there is supernatural causation of some natural events,

---

[26] Cf. David Papineau, who, though he takes naturalism to be committed to the completeness of physics, also maintains that although "a complete physics excludes psychology and that psychological antecedents are never needed to explain physical effects (...) psychological categories *can* be used to explain physical effects (...) in all such cases an alternative specification of a sufficient antecedent, which does not mention psychological categories, will also be available" (*op.cit.*, p. 31, n.26).

[27] It is natural to say 'tracing back to the creation' but we must remember that one kind of divine creation is "from eternity," in the sense that God creates the universe but not in time. This is perhaps arguably a kind of conserving causation; but, however it is to be conceived, I do not want to rule out the possibility. Note that determinism also implies an infinite past, on the assumption that there are events, that every event has an event cause, and that cause-effect relations take some minimal amount of time in the intuitive sense illustrated by paradigms of singular causal relations.

every mentally caused natural event *is* overdetermined. Why, however, should it matter to a scientific conception of the world that divine overdetermination—a kind of contemporaneous co-causation—could occur? So long as there are appropriate natural laws and natural antecedents of an event, it admits of scientific explanation under some description. One can leave open whether an explanation by appeal to events under mental descriptions is scientific.[28] No explanation adducing an event in the infinite chain of causes (as would be any causal explanation of an event) will be *ultimate*, but that should not be objectionable. Neither the attempt to reduce mentalistic explanations or properties to naturalistic ones nor the search for ultimate explanations is required in the scientific enterprise, and the latter seems characteristically forsworn.

Now consider the status of the scientific enterprise if the causal closure principle is true and every event is caused by another event. Then (on plausible assumptions about the temporality of causation) there will be an infinite series of events extending infinitely into the past. Moreover, no physical event, including the Big Bang (which I here assume occurred), escapes the need for—or at least the possibility of explanation in terms of—a further event. We must certainly find a causative or other explaining event if we are to understand the Big Bang nomically.

A different option is to countenance substance causation as an irreducible kind of causation, or to take causal closure to be simply the negative principle that the causes of natural events are never non-

---

[28] Cf. Richard Swinburne's view of overdetermination. He says (among other things) that when there is no connection between the joint determinants such that either one of them is dominant over the other(s) or they are common effects of the same cause, there is a coincidence. But he does not claim that there cannot be two distinct independent explanations. See *The Existence of God*, 2nd ed., Oxford University Press, Oxford 2004, p. 46. Overdetermination is indeed puzzling when the causal factors are taken to be independent; one naturally wants to say that the effect must be different. But consider instantaneous death in the case of a firing squad wherein all shots are deadly and simultaneous. The wound is different in some way given the multiple hits, but the death occurs at *t* as a result of each hit, and *it* is the significant event we take to be overdetermined.

-physical and then allow that some causes are themselves uncaused. We can then avoid the infinite past consequence. But now the cost of causal closure would be choosing between, on the one hand, positing some brute natural event, such as a big bang, so that we have no explanation of it, and, on the other hand, a kind of regress-stopping causation (substance causation) that contemporary naturalists (among others) have seldom if ever thought they understood. Naturalists tend to consider the idea that a substance simply causes an event, such as the Big Bang, at best deeply puzzling. And what *natural* substance could have such a momentous issue? One thinks here of an unmoved mover with power approaching that of the divine. But for naturalism, it cannot be God. Why, in any case, would either positing a brute natural event or substance causation be preferable for naturalism to countenancing an infinite regress? That regress, however, provides for no ultimate explanations and leaves the cosmological question (Why is there something and not nothing?) unanswered and—apparently—unanswerable? Even many naturalists may find the cost of the former alternatives too great.

Causal overdetermination

If all this is so, there is some reason to look again at the possibility of non-natural causal events overdetermining some natural events, or at least originatively producing them. In any case, it is the philosophical naturalist, not the scientist, who must resist the possibility of such overdetermination. The *causal sufficiency of the physical*—as a principle of plenitude taken to entail the existence of physical causally sufficient conditions for every natural event—implies the possibility of as comprehensive a physical science as we can reasonably develop and is consistent with a plausible version of methodological naturalism and even with some versions of philosophical naturalism. We can either rest content with a big bang or, in principle, search further for an adequate nomic account of the facts to be explained. But unlike the causal closure principle, the causal sufficiency view does not rule out supernatural causation.

Some philosophers will be uncomfortable with one implication of the sketch so far suggested as a possibility. If overdetermination of the kind in question is possible, we might have to allow the possibility that if, say, the physical cause of a natural event had not occurred, the event would still have occurred. But this need only be a logical possibility; there need be no nomically or causally possible world in which it occurs. Scientists depend on laws of nature, or at least on causal regularities, for their explanations in any case. Why must it disturb scientists, as opposed to certain philosophical naturalists, if, in other logically possible worlds (and these worlds may even be nomically or causally impossible), a phenomenon does not admit of the kind of explanations they seek?

I have in some places spoken as if determinism is true. But I have not presupposed it (and doubt that it is true). If it is not, then causal sufficiency does not require deterministic causes of all natural events. In that case, however, there may be a type of event which is of the right kind to cause a further event, but does not always do so. Suppose such a potentially causative event is present and, relative to the natural order, it is indeterminate whether the effect will occur. Its occurrence and its non-occurrence are both consistent with the laws of nature. If the situation is such that we could correctly explain the effect by appeal to the event if the event occurs, then we might also say that God's volitional action could occur in such a way as to "close the gap," but *without* either undermining or predetermining the connection between the non-deterministic cause and its effect. This is a presumptive case of overdetermination occurring in a non-deterministic causal chain in the natural world. Whether it occurs and, if so, how often, is a contingent matter. Determinists would of course reject the conception of nature that allows this phenomenon; but determinism is not a necessary truth, and (so far as I know) there are plausible theories in quantum mechanics that are commonly taken to provide reason to consider determinism false.

*   *   *

Philosophical naturalism can take many forms and is motivated in somewhat different ways by different theorists and in different sub-fields of philosophy. In its domain-specific forms, it is more plausible in some domains of inquiry than in others. Naturalization in the philosophy of mind, for instance, seems less difficult to defend than full-scale philosophical naturalism requiring a nominalistic account of all apparent presuppositions of the existence of abstract entities. Moreover, even within a single field, such as epistemology, some core notions, say knowledge, seem more amenable to naturalization than others, such as justification. In its strong forms, naturalization can be rationally resisted without extravagant metaphysical or epistemological claims. Moreover—and this is a more important conclusion for this paper—scientific rigor does not require a commitment to comprehensive naturalism, nor is any stronger naturalism than a methodological version required for cultivation and maintenance of a scientific habit of mind. We have also seen reason to think that a scientific worldview is neutral with respect to the causal closure of the world. Causal sufficiency would suffice as a metaphysical view that accommodates the ideal of unifying science under physical laws, though that ideal itself is not one that must be endorsed by inquirers in the special sciences. It appears, then, that neither science nor ethics nor logic and mathematics, nor philosophy itself, requires endorsing a full-scale philosophical naturalism for success in providing plausible theories of human experience.[29]

[29] An earlier version of this paper was given at the Jagiellonian University of Krakow as part of a series I presented on Naturalism, Normativity, and the Explanation of Human Action, and I benefited much from discussing it there. The paper also reflects much that I have learned from discussions with colleagues, friends, and audiences, and it draws in several places on previous work, most notably *Philosophical Naturalism at the Turn of the Century*, "Journal of Philosophical Research" 2000, vol. 25, pp. 27–45; *Can Normativity Be Naturalized?*, [in:] *Ethical Naturalism: Current Debates*, eds. S. Nuccetelli, G. Shea, Cambridge University Press, Cambridge 2012, pp. 169–193; and Chapter 10 of *Rationality and Religious Commitment*, Clarendon Press, Oxford 2011. Thanks to Mario De Caro for helpful comments on the penultimate draft.

Roman Murawski
Adam Mickiewicz University in Poznań

On Proof in Mathematics[1]

It is obvious nowadays that proof is the main method of justification in mathematics. All theorems should be proved, proof is the only method used and admissible in mathematics to convince others (the audience, readers) of the truth of a given statement. Only statements that have been proven can be treated as belonging to the realm of mathematics. Neither authority nor experiment are allowed in this process – at least theoretically. But what is a proof in mathematics? What does it mean to prove a theorem: to convince specialists in the given domain of mathematics of its truth or to give a proof in the sense of mathematical logic?

Observe that such a model of mathematics as a science was born only in ancient Greece in the 4th century B.C. Earlier, for example in ancient Egypt or Babylon, there were no proofs. In both of those pre--Greek mathematics – though they were advanced and sophisticated (especially Babylonian mathematics) – there was no need to prove statements. In fact there were no general statements and no attempts were undertaken to deduce the results or to explain their validity A similar case was that of Chinese mathematics.

---

[1] The publication of this paper was made possible through the support of a grant "The Limits of Scientific Explanation" from the John Templeton Foundation. The financial support of National Center for Science [Narodowe Centrum Nauki], grant No N N101 136940 is acknowledged.

Proofs as deduction from explicitly stated postulates were conceived of by the Greeks. It was connected with the axiomatic method. Since Plato, Aristotle and Euclid, the axiomatic method was considered as the best method to justify and to organize mathematical knowledge. The first mature and most representative example of its usage in mathematics were the *Elements* of Euclid. They established a pattern of a scientific theory and in particular a paradigm in mathematics. From Euclid until the end of the 19th century, mathematics was developed as an axiomatic – in fact rather a quasi-axiomatic – theory based on axioms, postulates and definitions. Axioms were principles common to all sciences, postulates – specific principles taken for granted by a mathematician engaged in the demonstration of theorems in a particular domain. Definitions should provide meaning to new notions – in practice, definitions were rather explanations of notions than proper definitions in the strict sense, moreover, they were explanations in the imprecise everyday colloquial language. Note that the language of a theory was not separated from the natural language. Proofs of theorems contained several gaps – in fact the lists of axioms and postulates were not complete, one freely used in proofs various "obvious" truths or referred to the intuition. Consequently, proofs were only partially based on axioms and postulates. In fact proofs were informal and intuitive, they were rather demonstrations and the very concept of a proof was of a psychological and sociological (and not of a logical) nature.

Note that almost no attention was paid to the precisation and specification of the language of theories – in fact the language of theories was simply the imprecise colloquial language. One should also note here that in fact until the end of the 19th century, mathematicians were convinced that axioms and postulates should be true statements and hence sentences describing the real state of affairs (in the mathematical reality). It seems to be connected with Aristotle's view that a proposition is demonstrated (proved to be true) by showing that it is a logical consequence of propositions already known to be true. Demonstration was conceived of here as a deduction whose premises are known to be true and a deduction was conceived of as a chaining of immediate inferences.

It should be noted that the Euclidean approach (connected with Platonic idealism) to the problem of the development of mathematics and the justification of its statements (which found its fulfilment in the Euclidean paradigm), i.e. justification by deduction (by proofs) from explicitly stated axioms and postulates, was not the only approach and method which was used in the ancient Greek (and later). The other one (call it heuristic) was connected with Democritean materialism. It was applied, for example, by Archimedes who used not only deduction but any methods, such as intuition or even experiments (not only mental ones) to solve problems. One can see this, for example, in his considerations concerning the calculation of the volume of a sphere using cylinder with two excavated cones or in his quadrature of the parabola.

Although the Euclidean approach won and dominated in historical terms, one should note that it formed rather an ideal than the real scientific practice of mathematicians. In fact, rigorous, deductive mathematics was a somewhat rare phenomenon. On the contrary, intuition and heuristic reasoning were the animating forces of mathematical research practice. The vigorous but rarely rigorous mathematical activity produced "crises" (for example the Pythagoreans' discovery of the incommensurability of the diagonal and the side of a square, Leibniz's and Newton's problems with the explanation of the nature of infinitesimals, Fourier's "proof" that any function is representable in a Fourier series, antinomies connected with Cantor's imprecise and intuitive notion of a set).

New elements appeared in the 19th century with the trend whose aim was the clarification of basic mathematical concepts, especially those of analysis (cf. works by Cauchy, Weierstrass, Bolzano, Dedekind). Yet another factor was the discovery of antinomies, in particular in set theory (C. Burali-Forte, G. Cantor, B. Russell) and of the semantical antinomies (G.D. Berry, K. Grelling). They forced the revision of some basic ideas of mathematics. Among formulated proposals was the foundational program of David Hilbert and his *Beweistheorie*. Note that "this program was never intended as a comprehensive philosophy of mathematics; its purpose was instead to legitimate the entire corpus of mathematical knowledge."[2]

The main role in Hilbert's program was played by formal (or formalized) proofs. This device was introduced on the basis of (and thanks to) the mathematical logic developed in the 19th century, in particular on the basis of the work of Gottlob Frege who constructed the first formalized system – it was the system of propositional calculus based on two connectives: negation and implication. Investigations carried out in the framework of Hilbert's program established a new scientific discipline, i.e., metamathematics.

Recall that the concept of a formal proof must be related to a given formal theory. To express such a theory one should at the beginning fix its language. The rules of forming formulas in it should have strictly formal and syntactic character – only the shape of symbols can be taken into account and one should entirely abstract from their possible meaning or interpretation. Next one fixes axioms (logical, non-logical and usually identity axioms) and rules of inference. The latter must have entirely syntactic and formal character. A formal proof of a statement (formula) φ is now a finite sequence of formulas in the given language $\varphi_1, \varphi_2, \ldots, \varphi_n$ such that the last member of the sequence is the formula φ and all members of it either belong to the set of presumed axioms or are consequences of previous members of the sequence according to one of the accepted rules of inference. Observe that this concept of a formal proof has a syntactic character and does not refer to any semantical notions such as meaning or interpretation.

As a result of the described development, one nowadays in mathematics has at least two concepts of a proof: the formal one, used mainly by logicians and specialists in metamathematics in the foundational studies and on the other hand the "normal", "usual" concept of a proof used in mathematical research practice. What are the rela-

---

[2] Cf. D.E. Rowe, *Klein, Hilbert, and the Göttingen Mathematical Tradition*, "Osiris" 1989, vol. (2)5, p. 200.

tions between them? Is the metamathematical concept a precisation of the "everyday" concept? Do they play similar roles in mathematics or not? To what extent does the concept of a formal proof reflect the main features and the nature of an informal proof used by mathematicians in their researches?

Note at the beginning that formal proofs did not replace informal ones – the latter did not disappear from mathematical practice. In fact mathematics was and is developed in an informal way using intuition and heuristic reasonings – it is still developed in fact in the spirit of Euclid (or sometimes of Archimedes) in a quasi-axiomatic way. Moreover, informal reasonings appear not only in the context of discovery but also in the context of justification. Any correct methods are allowed to justify statements. But what does it mean "correct"? In the research practice this is decided by the community of mathematicians. The criteria of being correct were changing in the history and in the process of developing mathematics. The main aim of a mathematician is always to convince the audience that the given result is justified, correct and true (and this concept is understood in an intuitive and vague way) and not to answer the question whether it can be deduced from stated axioms. The ultimate aim of mathematics is "to provide correct proofs of true theorems."[3] In mathematical research practice, the concepts "true" and "provable" are usually not distinguished and often replaced by each other. Mathematicians used to say that a given theorem holds or that it is true and not that it is provable in such and such theory. Axioms of developed theories are not always precisely formulated and admissible methods precisely described.

Proofs play various roles in the mathematical research practice. One can distinguish among others the role of (1) verification, (2) explanation, (3) systematization, (4) discovery, (5) intellectual challenge, (6) communication, (7) justification of definitions.[4]

---

[3] J. Avigad, *Mathematical Method and Proof*, "Synthese" 2006, vol. 153, p. 105.

[4] Cf. M.D. De Villiers, *Rethinking Proof with the Geometer's Sketchpad*, Key Curriculum Press, Emeryville, CA. 1999; T. Cadwallader Olsker, *What Do We Mean by Mathematical Proof?*, "Journal of Humanistic Mathematics" 2011, vol. 1, pp. 33–60.

The most familiar to research mathematicians is the role of verification – a statement can be treated as belonging to the body of mathematics only when it has been verified. The proof should not only show that a given sentence is true and holds but should also explain why it is true and holds. This role explains why mathematicians are often looking for new proofs of known theorems – new proofs should have more explanatory power. The role of systematization was exemplified already by Euclid's *Elements*. In this work, many theorems known to Greeks have been collected and organized in such a way that they followed from axioms, postulates, definitions and previously proved theorems. It was shown in this way that the accepted axioms, postulates and definitions form a sufficient base on which the whole edifice of mathematics can be developed. Note that the role of discovery may be – *prima facie* – rather seldom associated with proofs but it is not excluded. In fact, e.g., non-Euclidean geometries were arrived at through purely deductive means. Recall that since Euclid, one asked the question whether the fifth postulate on parallels formulated in *Elements* is independent of other axioms and postulates or can be deduced from them. After several attempts undertaken through the centuries, it was shown in the 19th century that it is reasonable to consider systems of geometry in which the negation of the fifth postulate is assumed instead of the fifth postulate itself and it is possible to show that such systems are consistent.

Finding proofs is an intellectual challenge for mathematicians: there is a theorem – we want to prove it. Proofs serve in the community of mathematicians as communication means. They communicate not only the reasons why a given statement is a true theorem but also new methods which can be used sometimes in other domains. Proofs can also provide justification of definitions.

The most important roles played by proofs in research practice are, of course, verification and explanation. Note that a proof that verifies a theorem does not have to explain why it holds. One can distinguish between proofs that convince (that a statement holds or is true and can be accepted) and proofs that explain (why it is so). Of course, there are proofs that both convince and explain. The explanatory proof should give an insight in the matter whereas the convincing one should be concise or general. One can distinguish also between explanation and understanding. Mathematicians often treat simplicity as a characteristic feature of understanding. Observe that, as G.-C. Rota writes, "[i]t is an article of faith among mathematicians that after a new theorem is discovered, other, simpler proof of it will be given until a definitive proof is found."[5] There are many examples from the history of mathematics that confirm this. In this context one can mention Paul Erdős' idea of proofs from The Book in which God maintains the perfect proofs for mathematical theorems – he followed the dictum of G.H. Hardy that there is no permanent place for ugly mathematics. Erdős stressed that one need not believe in God but, as a mathematician, one should believe in The Book.[6]

Note that a proof that convinces can be more (or even quite) formal. Explanatory proofs cannot be strictly formal. Mathematicians set a high value on explanatory proofs. Such a proof is more valued when "it exhibits methods that are powerful and informative."[7] Hersh[8] says that "[p]roof is a tool in service of research, not a shackle on the mathematician's imagination."[9]

One can verify a theorem by checking all particular cases but this usually does not give an explanation why the theorem holds. The

[5] G.-C. Rota, *The Phenomenology of Mathematical Proof*, "Synthese" 1997, vol. 111, p. 192.

[6] In 1998 Springer Verlag published *Proofs From The Book* by M. Aigner and G.M. Ziegler. It is treated by the authors as "a first (and very modest) approximation of The Book" (Preface).

[7] Cf. J. Avigad, *Mathematical Method and Proof*, "Synthese" 2006, vol. 153, p. 106.

[8] R. Hersh, *What Is Mathematics, Really?*, Oxford University Press, New York 1997, p. 60.

[9] On the concept of explanation in mathematics see for example M. Steiner, *Mathematical Explanation*, "Philosophical Studies" 1978, vol. 34, pp. 133–151, P. Mancosu, *On Mathematical Explanation*, [in:] *The Growth of Mathematical Knowledge*, eds. E. Grosholz, H. Breger, Kluwer Academic Publishers, Dordrecht 2000, pp. 103–119, P. Mancosu, *Mathematical Explanation: Problems and Prospects*, "Topoi" 2001, no. 20, pp. 97–117.

explanation should give a general principle by which the theorem holds. A famous example of a theorem verified by checking cases but not giving reasons is the Four-Color Theorem proved by Appel and Haken and stating that every planar graph is four-colorable, i.e., in another words, that four colours suffice to colour every map on the plane in such a way that two regions receive different colours whenever they have a common border. The example of this theorem indicates other features of proofs in mathematics. Observe that the first purported proof of it given by Kempe in 1879 was accepted for a decade before it was found to be incorrect. This was neither the first nor the last example of such a situation. It means that the community of mathematicians can be fallible.

The Four-Color Theorem opens our eyes to the problem of methods acceptable in a proof or in a verification of cases. The unique known proof – that by Appel and Haken – was obtained by using computer and no traditional proof (without a computer) is known so far. Moreover, the existing proof cannot be made by a human being because an essential part of it was a computation requiring about 1200 hours of computer time and which is beyond the capacities of any mathematician. This initiated a discussion concerning the admissibility of experimental methods in mathematical proofs. Several arguments for and against have been formulated – we will not enter here the details of the discussion. Let us say only that the usage of a technical tool (like a computer) seems to refute the commonly accepted thesis that mathematical knowledge is *a priori*. There also arises a question whether a computer-aided proof is (or can be treated as) a mathematical proof and consequently, whether in particular the Four-Color Theorem can be called "theorem" or rather still a hypothesis.

In this debate, initiated by a paper by T. Tymoczko,[10] the question was asked as to what are in fact the characteristic features of a "nor-

---

[10] T. Tymoczko, *The Four-Color Problem and Its Philosophical Significance*, "The Journal of Philosophy" 1979, vol. 76, pp. 57–83.

mal" mathematical proof. Tymoczko says that a proof in mathematics should be: (1) convincing, (2) surveyable and (3) formalizable. The first feature is of an "anthropological" character (as Tymoczko says), the other two he treats as "deep features". He claims also that "surveyability and formalizability can be seen as two sides of the same coin."[11] Formalizability "idealizes surveyability, analyzes it into finite reiterations of surveyable patters."[12] It can be assumed that all surveyable proofs are formalizable. Are also all formalizable proofs surveyable? Tymoczko answers this negatively: "[w]e know that there must exist formal proofs that cannot be surveyed by mathematicians if only because the proofs are too long or involve formulas that are too long" and the phrase "too long" means here "can't be read over by a mathematician in a human life time."[13] On the other hand one should observe that "it is not at all obvious that mathematicians could come across formal proofs and recognize them as such without being able to survey them."[14]

Considering surveyability, one should distinguish local and global surveyability. Bassler characterizes them in the following way: "local surveyability requires the surveying of each of the individual steps in a proof in some order, while global surveyability requires the surveying of the entire proof as a comprehensive whole."[15] Hence a local surveyability does not mean that a proof is practically surveyable. One can say that the proof of the Four-Color Theorem is globally surveyable without being locally surveyable (provided one is willing to countenance a distinction between proof and calculation). On the other hand, if one accepts the assumption that global surveyability receives its foundation in local surveyability then this statement is false. Add that one should also distinguish the surveyability of a proof and

---

[11] *Ibid.*, p. 61.
[12] *Ibid.*
[13] *Ibid.*
[14] *Ibid.*, p. 62.
[15] G.B. Bassler, *The Surveyability of Mathematical Proof: A Historical Perspective*, "Synthese" 2006, vol. 148, p. 100.

the fact that it can be (formally) checked on the one hand and the fact that it gives an understanding, that it reveals the deep reasons for the theorem being proved on the other.

The concept of surveyability is not precise enough. In the 20th century, there was a trend to link surveyability with the development of formal and complete foundations of mathematics and formalization was treated as a method providing the local surveyability. The works of Frege, Russell and their followers, especially Hilbert, were guided by the desire to find a perspicuous syntactic representation of the relations of semantic content within a proposition.

Computers, and the methods connected with them, were and are used in mathematics not only in the proof of the Four-Color Theorem. They are used in various contexts in mathematics, in particular (1) to perform numerical calculations, (2) to find (usually approximate) solutions of equations and systems of algebraic or differential equations or of integrals, (3) in automating proofs of theorems, (4) in checking the correctness of mathematical proofs, (5) in proving theorems (one says then about computer-aided proofs) and (6) in various experiments with mathematical objects (e.g., in the theory of fractals). From our point of view the most important applications are (3) and (4) – the application of the type (5) has been discussed above on the example of the Four-Color Theorem.

The automating proving of theorems is connected with the idea of mechanization and automatization of reasoning due to Leibniz.[16] This idea (as one of the factors) led to the development of formal logic and in consequence to the idea of a formal proof.

Formal proofs were introduced to provide an explication of the informal notion of a proof. They should explain the virtue by which the usual proofs used in the research practice are judged to be correct. They should also explain what does it mean that a given statement is a logical or deductive consequence of certain assumptions. They were

---

[16] Cf. W. Marciszewski, R. Murawski, *Mechanization of Reasoning in a Historical Perspective*, Editions Rodopi, Amsterdam/Atlanta, GA 1995.

introduced in the atmosphere of a crisis in the foundations of mathematics. For Frege and Russell, they were means to an end, a way of precisely isolating the permissible proofs and making sure that all use of axioms was explicit. On the other hand, observe that Hilbert was not really interested in actually formalizing proofs and replacing the "normal" research proofs by formalized ones. He treated formalization and formal proofs as a tool to justify mathematics as a science and to establish its consistency. They should serve theoretical purposes – in particular to prove results about mathematics, hence to obtain metamathematical results.

In fact, it was the development of logic and the concept of a formal proof based only on syntactical properties which made the development of metamathematics possible. A lot of interesting results have been obtained here. First of all, the old paradigm of mathematics that was functioning since Euclid has been made precise – in fact it has been replaced by a new logico-settheoretical paradigm.[17] The main features of this new paradigm can be described as follows: (1) set theory became the fundamental domain of mathematics, in particular some set-theoretical notions and methods are present in any mathematical theory and set theory is the basis of mathematics in the sense that all mathematical notions can be defined by primitive notions of set theory and all theorems of mathematics can be deduced from axioms of set theory, (2) languages of mathematical theories are strictly separated from the natural language, they are artificial languages and the meaning of their terms is described exclusively by axioms; some primitive concepts are distinguished and all other notions are defined in terms of them according to precise rules of defining notions, (3) mathematical theories have been axiomatized,(4) there is a precise and strict distinction between a mathematical theory and its language on the one hand and metatheory and its metalanguage on the other (the distinction was explicitly made by A. Tarski). Note also that two

---

[17] Cf. R. Murawski, *Dwa paradygmaty matematyki. Studium z dziejów i filozofii matematyki*, Wydawnictwo Naukowe Uniwersytetu im. Adama Mickiewicza, Poznań 1996.

concepts, crucial for mathematics: the concept of a syntactical consequence (being provable) and the concept of being a semantical consequence have been precisely defined and strictly distinguished. One
could also precisely distinguish provability and truth. In a "normal"
research mathematical practice – as we indicated above – they are
usually identified or at least not distinguished – one says that a theorem holds, i.e., has been proved, or that it is true and treats both as
equivalent and synonymous. The very process of distinguishing them
was long and not so simple.[18] The crucial role was played here by
Gödel's incompleteness theorems.

The incompleteness results of Gödel showed that there exist
sentences that are true but are neither provable nor refutable, i.e.,
that are undecidable in a given theory. Before Gödel it was believed
that formal demonstrability is an analysis of the concept of mathematical truth. Gödel wrote in a letter dated 7th March 1968 to Hao
Wang:

> (...) formalists considered formal demonstrability to be an *analysis* of
> the concept of mathematical truth and, therefore were of course not in
> a position to *distinguish* the two.[19]

Indeed, the informal concept of truth was not commonly accepted as
a definite mathematical notion at that time.[20] Gödel wrote in a crossed-
-out passage of a draft of his reply to a letter of the student Yossef Ba-

---

[18] Cf. R. Murawski, *Truth vs. Provability—Philosophical and Historical Remarks*,
"Logic and Logical Philosophy" 2002, vol. 10, pp. 93–117. Reprinted in: R. Murawski,
Essays in the Philosophy and History of Logic and Mathematics, Editions Rodopi,
Amsterdam/New York, NY, pp. 41–57; R. Murawski, *On the Distinction Proof-Truth in
Mathematics*, [in:] *In the Scope of Logic, Methodology and Philosophy of Science*, eds.
P. Gärdenfors et al., Kluwer Academic Publishers, Dordrecht–Boston–London 2002,
pp. 287–303.
[19] H. Wang, *From Mathematics to Philosophy*, Routledge and Kegan Paul, London
1974, p. 10.
[20] Note that there was (at that time) no precise definition of truth — this was given
in 1933 by A. Tarski (cf. A. Tarski, *Pojęcie prawdy w językach nauk dedukcyjnych*,
Nakładem Towarzystwa Naukowego Warszawskiego, Warszawa 1933).

las: "(...) a concept of objective mathematical truth as opposed to demonstrability was viewed with greatest suspicion and widely rejected as meaningless."[21] It is worth comparing this with a remark of R. Carnap. He wrote in his diary that when he invited A. Tarski to speak on the concept of truth at the September 1935 International Congress for Scientific Philosophy, "Tarski was very sceptical. He thought that most philosophers, even those working in modern logic, would be not only indifferent, but hostile to the explication of the concept of truth." And indeed at the Congress "(...) there was vehement opposition even on the side of our philosophical friends."[22]

All this explains in some sense why Hilbert preferred to deal in his metamathematics solely with the forms of the formulas, using only finitary reasonings which were considered to be safe – contrary to semantical reasonings which were non-finitary and consequently not safe. Non-finitary reasonings in mathematics were considered to be meaningful only to the extent to which they could be interpreted or justified in terms of finitary metamathematics.[23]

On the other hand, there was no clear distinction between syntax and semantics at that time. Recall, for example, that the axiom systems came by Hilbert often with a built in interpretation. Add also that the very notions necessary to formulate properly the different syntax–semantics were not available to Hilbert.

Gödel proved that truth cannot be adequately achieved and expressed by provability, that the whole of mathematics (or even parts of it) cannot be included in a formalized system. This indicated certain weakness of the concept of a formal proof. Gödel's results showed also that one should not limit or bound the creative invention of mathematicians. In the framework of formalized theories one can extend

---

[21] Cf. H. Wang, *Reflections on Kurt Gödel*, MIT Press, Cambridge, Mass. 1987, pp. 84–85.
[22] R. Carnap, *Intellectual Autobiography*, [in:] *The Philosophy of Rudolf Carnap*, ed. P.A. Schilpp, Open Court Publishing Co., La Salle, Ill., 1963, pp. 61–62.
[23] Cf. Gödel's letter to Hao Wang dated 7th December 1967 — see H. Wang, *From Mathematics...*, *op.cit.*, p. 8.

them by adding new axioms or by admitting new inference rules. The second possibility means that infinitary rules are admitted – but this changes the whole picture and the whole paradigm! On the other hand, one can ask whether the process of adding new axioms, though necessary to solve problems which are not decidable in a given theory, is sufficient? Will it suffice to express the creativity of a mathematical mind, the creativity of mathematicians?

The incompleteness theorems of Gödel belong to the so-called limitation results. They are results stating that certain properties important and desired from a metamathematical point of view (but also from the point of view of a working mathematician) cannot be achieved. Among them are the theorem of Tarski's, which states the undefinability of the concept of truth, and the theorem of Löwenheim and Skolem, showing that a mathematical structure cannot be adequately and uniquely described by a formalized theory (a theory having a model has in fact many various models). Tarski wrote:

> It was undoubtedly a great achievement of modern logic to have replaced the old psychological notion of proof, which could hardly ever be made clear and precise, by a new simple notion of a purely formal character. But the triumph of the method carried with it the germ of a future setback.[24]

Considerations concerning formal proofs also enlightened the role played by infinity in mathematics, in particular in the process of proving. Gödel's results show that finite/finitistic methods of syntactical formal provability do not exhaust the variety of mathematical truth. In fact, if one wants to obtain a complete theory then some infinite/infinitistic rules (such as, e.g., the ω-rule) are necessary. Recall that the ω-rule is an inference rule with infinitely many premises, i.e., it is the following rule:

---

[24] Cf. A. Tarski, *Truth and Proof*, "Scientific American" 1969, vol. 220(6), p. 74.

$$\frac{\varphi(1), \varphi(2), ..., \varphi(n), ... \ (n \in N)}{\forall x \varphi(x)}$$

In mathematical research practice nobody restricts himself/herself to finite methods, on the contrary, any correct methods, among them infinite (in particular set-theoretical and semantical) are applied.

Being not as ideal as it was hoped, formal proofs play an important role in metamathematics, i.e., in the study of mathematical theories or of mathematics as a collection of theories – but not only there. They enable also the automatization and mechanization of proofs in mathematics, hence they make possible the construction of automated proofs and the verification of proofs by computers. Verification of (formal) proofs is possible because the relation "x is a proof of y" is – as was shown in mathematical logic – recursive, hence effective. On the other hand, the construction and finding of proofs is not an effective procedure (is not recursive), in fact it is only recursively enumerable. Therefore one can say that "[f]ormalization is about checking, and not about discovery."[25] The results of Turing, Church and Gödel have shown that there is no universal method of finding or constructing (formal) proofs. Hence every formal proof is a result of a creative invention of a human being.

Observe that the concept of a formalized proof is one for all theories, it is in a sense a uniform concept. It is independent of subjective, cultural and sociological elements and factors. Moreover, the completeness theorem of Gödel states that the logical means of the first-order logic (first-order predicate calculus) are sufficient (in the case of, e.g., second-order logic the situation is different – one does not have here the completeness phenomenon). This concept enables us to make the concept of a proof more objective. It also makes possible the precise study of provability in mathematics (under the assumption that the logical concept of a proof reflects all important and

---

[25] Cf. F. Wiedijk, *Formal Proof – Getting Started*, "Notices of the American Mathematical Society" 2008, vol. 55, p. 1414.

essential features of proofs from the research practice of mathematicians). One can prove results stating that a given statement is not a theorem of a given theory, i.e., that there exist no proof of a given statement or that a given sentence is undecidable (in a given theory). It also enables the study of important properties of mathematical theories such as consistency, completeness, independence of axioms or axiomatizability in a given way, etc.

The concept of a formal proof also serves the philosophy of mathematics. It helps in attempts to answer the question about the existence and character of mathematical objects as well as in considerations concerning the epistemology of mathematics. On the other hand all philosophies of mathematics reducing mathematics to formalized axiomatic theories (among them logicism and formalism) have a reductionist character and do not take into account the actual research practice of mathematicians. Their aim is to justify mathematics and not to explain real mathematical practice. It is worth noting that in recent trends in the philosophy of mathematics still more and more attention is paid to the study just of the research practice in mathematics – one takes into account various sociological, psychological and cultural factors. Unfortunately it is done only by the analysis of particular discoveries and achievements, hence by case studies. There are no general conceptions. But is it possible to develop such general conceptions?

In fact formal proofs are connected rather with foundational studies than with research practice. Observe that a formal proof does not give an understanding, it does not explain the deep reasons of a theorem. They are also not suitable for the practice – they are simply too long, they are too tedious and painstaking. In such a proof the underlying intuition may get lost. Formalized mathematics may be also more error-prone than the usual informal one – in fact formal manipulations may become very complicated. As Bourbaki wrote:

> If formalized mathematics were as simple as the game of chess, then once our chosen formalized language had been described there

would remain only the task of writing out proofs in this langauge, (…) But the matter is far from being as simple as that, and no great experience is necessary to perceive that such a project is absolutely unrealizable: the tinest proof at the beginning of the Theory of Sets would already require several hundreds of signs for its complete formalization. (…) formalized mathematics cannot in practice be written down in full, (…). We shall therefore very quickly abandon formalized mathematics, (…).[26]

Cadwallader-Olsker writes:

A purely formal proof (…) cannot be very complex without becoming so lengthy as to be incomprehensible to a human reader. Such a formal proof is rarely able to be explanatory, and may only be convincing to the degree that it can be read and understood by the reader or checked by a computer.[27]

Add that a transcription of a single traditional (hence informal) proof into a formal one is a major undertaking.

*   *   *

We have shown that one has two concepts of a proof in mathematics: an informal one used by mathematicians in their usual research practice and the concept of a formal or formalized proof used mainly in logic and the foundations of mathematics. The first one is not defined precisely, it is simply practised and any attempts to define it fail. It is – so to speak – a practical notion. It has a psychological, sociological and cultural character. The second one is precisely defined in terms of logical concepts. Hence it is a logical concept having rather theoretical

---

[26] Cf. N. Bourbaki, *Theory of Sets*, Addison-Wesley Publishing Company, Reading, Mass. 1968.
[27] T. Cadwallader-Olsker, *What Do We Mean by Mathematical Proof?*, "Journal of Humanistic Mathematics" 2011, vol. 1, p. 42.

than practical character. The first one has – in a part at least – semantical character, the second is entirely syntactical in nature.

Hence this situation can be compared with the situation concerning the Church-Turing thesis. This thesis states the equivalence of two concept: effective computability (in the intuitive sense) and recursivity (or Turing computability or computability in the sense of Markov or any other precisely defined and equivalent sense). As is well known, this equivalence cannot be proved with the degree of precision usual and required in mathematics because one part of it contains an intuitive vague concept formulated in the everyday language and the other a precise concept defined in the language of mathematics.[28] With such a situation one has to do also in other parts of mathematics – see for example the concepts of function, of truth, of logical validity or of limit.[29] In fact, until the 19th century, a function was tied to a rule for calculating it, generally by means of a formula. In the 19th and 20th centuries, mathematicians started to define a function as a set of ordered pairs satisfying appropriate conditions. The identification of those notions, i.e., of an intuitive notion and the precise set-theoretical one, can be called "Peano thesis". Similarly "Tarki's thesis" is the thesis identifying the intuitive notion of truth and the precise notion of truth given by Tarski. The intuitive notion of a limit, widely used in mathematical analysis in the 18th century and then in the 19th century, was applied by A. Cauchy to define basic notions of the calculus and was only given a precise form by K. Weierstrass in the language of $\varepsilon - \delta$. There are many other such examples: the notion of a measure as an explication of area and volume, the definition of dimension in topology, the definition of velocity as a derivative, etc.

---

[28] Cf. R. Murawski, *Church's Thesis and Its Epistemological Status*, "Annales Universitatis Mariae Curie-Skłodowska, Sectio AI, Informatica" 2004, vol. 2, pp. 57–70. Reprinted in: R. Murawski, *Essays in the Philosophy and History of Logic and Mathematics*, Editions Rodopi, Amsterdam/New York, NY, pp. 123–134; R. Murawski, J. Woleński, *The status of Church's Thesis*, [in:] *Church's Thesis After 70 Years*, eds. A. Olszewski, J. Woleński, R. Janusz, Ontos Verlag, Frankfurt 2006, pp. 310–330.

[29] Cf. E. Mendelson, *Second Thoughts about Church's Thesis and Mathematical Proofs*, "The Journal of Philosophy" 1990, vol. 87, pp. 225–233.

Comparing two concepts of a proof in mathematics, one can formulate a thesis stating that they are equivalent – one can call such a thesis a proof-theoretical thesis. As in the case of Church Thesis, no precise and strict proof of it can be given. One can only formulate arguments in favour or against it. The main argument for this thesis is the popular (among mathematicians and logicians) conviction that every "normal" mathematical proof can be formalized, i.e., can be written as a formal proof in a suitable axiomatic theory. There are, of course, no general rules regarding how to do this. A formalization of a proof requires often some original and not so obvious idea.

In this way we come to the following conclusion: there are two concepts of a proof in mathematics. They play different but complementary roles: formal proofs are used mainly in metamathematical and logical considerations, whereas informal proofs are used in the research practice of mathematicians.

Krzysztof Wójtowicz
University of Warsaw

Logical Form and Ontological Commitments[1]

Introduction

In this article I will indicate the problem of the interplay between ontology and "ideology": identifying the ontological commitments of theories depend on the assumptions concerning the semantic resources. I consider the example of the Boolos quantifier (but this particular choice is not essential) and formulate some remarks concerning the debate concerning the ontological commitments of mathematical theories.

1. The Problem of Ontological Commitments

It is quite clear that in our everyday communication we pay little attention to the problem of the underlying ontology. We simply say "I've got a headache" without analyzing the problem of the existence of a headache (as a process or an event or a property etc). In many simple cases (as this particular one), we can give paraphrases (e.g. we can simply say "My head is aching" instead of "I have got a headache") which allow us to "explain away" the allegedly existing objects.

---

[1] The publication of this paper was made possible through the support of a grant "The Limits of Scientific Explanation" from the John Templeton Foundation.

This model example shows, that – depending on the form of the utterance – we can identify the underlying ontology differently, and that the form of our statements does not always allow us to determine what are its existential presuppositions. We often claim that a certain object exists in a weak or metaphorical sense etc; sometimes we use some expressions in an elliptical way etc. So the problem of identifying existential commitments is not trivial and becomes particularly interesting in the case of scientific theories. In this case, it is necessary to give a criterion for determining whether a given theory really assumes the existence of objects of a certain type. One possible answer is provided by Quine's famous criterion of existence (the quantifiability criterion), according to which "to be is to be a value if a variable." The concept of ontological commitment is, in this case, relative to a given theory, without having an absolute character (i.e. we rather speak of the existence relative to a given theory T, than existence *simpliciter*).[2] According to Quine, we are committed to those entities which have to appear in the range of bounded variables of existential sentences of a theory. We can therefore identify the ontological commitments of a theory in a simple way: by translating it into first order logic and identifying the existential sentences, which follow from the theory. Of course, according to Quine, the real meaning (and in particular the real ontological commitments) can be discovered by transforming the sentence into "canonical", i.e. first-order notation.

Quine's criterion rests on some assumptions which might be labelled as "ideological." In particular, Quine's criterion rests on the so called *first-order thesis*, according to which the real logic is first-order logic (i.e. first-order predicate calculus) with its semantics etc. The problem of the nature of logic has been discussed extensively, and I will not discuss the problem here. Nevertheless, I would like to stress that limiting our investigations only to elementary logic might be considered restrictive: it reduces the possibilities available to un-

---

[2] Cf. e.g. W.V.O. Quine, *Existence and Quantification*, [in:] *idem*, *Ontological Relativity and Other Essays*, Columbia University Press, New York 1969, pp. 91–113.

derstand a given concept and the role it plays in our scientific theories. This limitation can hardly be justified by scientific practice, as a scientist will accept any language or theory which will make his theory effective. In particular, it is highly improbable that the scientist would restrict himself to elementary language because of the first-order thesis, which depends only on philosophical assumptions concerning the status and nature of logic. I think, that this is also the case of mathematics: mathematicians do not bother to formulate their theories in first-order (and not second-order) languages, but accept the arguments which they consider convincing and satisfactory. First-order logic is not distinguished in any way – the only aspect that matters is the possibility of describing our object of study.

2. The Boolos Quantifiers

As an example, I would rather indicate some problems concerned with the first-order thesis by discussing the problem of logics with additional quantifiers. Logics with additional quantifiers have been discussed very extensively,[3] in particular in connection with the linguistic problems of natural language. A very interesting example of a situation, where a certain "ideological decision" has to be made, is the case of the plural quantifier investigated by Boolos.[4] The natural language representation of the plural quantifier is the expression "There are...", which are used in sentences like "There are chairs in this room." In this case, the formalization in elementary logic is straightforward:

$$\exists x \exists y (C(x) \wedge C(y) \wedge R(x) \wedge R(y) \wedge x \neq y)$$

---

[3] Cardinality quantifiers, branching quantifiers etc.

[4] G. Boolos, *To Be Is To Be a Value of a Variable (or to Be Some Values of Some Variables)*, "Journal of Philosophy" 1984, vol. 81, pp. 430–450, (reprinted in: G. Boolos, *Logic, Logic, and Logic*, Harvard University Press, Cambridge, Mass. 1998).

Being precise, this sentence exactly formalizes the statement "There are at least two chairs in this room." But it is not always possible to formalize a sentence with "there are…", as the famous Geach-Kaplan example shows. Consider the sentence *"Some critics admire only one another."* There is no first-order formal paraphrase of this sentence. The question arises, how this sentence should be "deciphered". According to Quine, the appropriate formalization (paraphrase) is in the form of a second-order sentence of the form *"There is a certain non-empty* set *of critics, such that…"*, formally:

$$\exists S (\exists x\, x \in S \land \forall u \forall v [(u \in S \land Puv) \to (v \in S \land u \neq v)])$$

This sentence makes use of second-order quantification. Therefore, the question arises as to what the ontological commitments of the Geach-Kaplan sentence are. If we adopt the orthodox Quinean point of view, we should agree, that some ontological commitments to the existence of sets are involved (as the formal paraphrase uses the second-order quantification). We could claim, that in the GK-sentence, ontological commitments to sets are hidden, which are revealed by transposing the sentence into second-order form (as there is no first-order paraphrase).

However, it is not obvious that this is the proper paraphrase. Boolos proposes to treat the plural quantifier *"There are such x, that…"* as perfectly legitimate and intelligible without second-order translation. We use such expressions in our natural language communication, and they seem to be quite clear without translating them. Formally, we use the notation $\exists xx$ for *"there are x, such that"* (observe, that first order variables are bounded here, so this quantifier can be treated as first-order quantifier!) and the relation *"…to be one of…"*. I will use the symbol $\langle$, so $a\langle xx$ means "a is one of the xx." The relation $\langle$ obtains between singular and plural terms, so we can formalize the Geach-Kaplan sentence in the following way:

$$\exists xx\, \forall u\, \forall v\, [(u\langle xx \to Cu) \,\&\, (Auv \to v\langle xx \land u \neq v)].$$

We interpret the plural quantifier as binding individual variables – so this is not a kind of second-order quantification.

The logic with plural quantification has a greater expressive power than first-order logic. Boolos[5] considers the sentence

> There are some horses that are all faster than Zev and also faster than
> the sire of any horse that is slower than all of them

(which can be also given an arithmetical counterpart),[6] which cannot be symbolized in first order logic. But this sentence characterizes exactly the non-standard models of arithmetic, so it clearly cannot have a first-order counterpart.

3. Ontology, Ideology and Ontological Commitments

The question arises as to whether we should treat sentences with "There are…" as autonomous and understandable without introducing set-theoretic paraphrases, or whether we should rather consider them to be notational variants of the "genuine" form (given by the second-order formalization). According to the first-order thesis, we should translate the sentences with plural quantifiers (in particular the GK-sentence) into the first-order sentence (and if this is not possible – into second order sentence). According to the more liberal view – this is not necessary, as plural quantifiers are perfectly legitimate.

The decision we make has an obvious influence on the discussion concerning ontological commitments. If we accept plural quantifiers as primitive, we can also proceed in the opposite direction, formalizing (some of the) second-order sentences as sentences with plural

---

[5] G. Boolos, *To Be Is To Be…*, *op.cit.*; G. Boolos, *Nominalist Platonism*, "Philosophical Review" 1985, vol. 94, pp. 327–44, (reprinted in: G. Boolos, *Logic, Logic, and Logic*, Harvard University Press, Cambridge, Mass. 1998).

[6] E.g. "There are such numbers, that are all greater than 0, and the successor of any n, which is smaller than all of them, is also smaller than all of them."

quantifiers – eliminating certain ontological commitments in this way (provided, of course, that we accept the Quinean view concerning the identification of ontological commitments).

The problem of expressive power therefore becomes important here. By introducing new quantifiers we can increase the expressive power without increasing ontological commitments. This is, for example, the point of view represented by Hellman. He accepts the thesis that plural quantifiers are can be considered to be legitimate semantic notions as they stand, without invoking set-theoretic translations.[7] Hellman's considers plural quantification when applied to sets of real numbers – it has the same expressive power as ordinary quantification applied to **families** of sets of real numbers (or in other words – to real functions and relations). This in particular means that, if we accept the Boolos quantifiers, we will be able to have the expressive power or RA^3 (third-order arithmetics) while restricting the ontological commitments to those of RA^2 (second-order arithmetics). Hellman claims, that in principle applied mathematics can be reconstructed within RA^3, which means in particular, that the ontological commitments on applied mathematics are limited to those of RA^2. Hellman also claims that it is possible to provide a nominalistically accepted reconstruction of RA^2. I will not go into technical details of Hellman's program here,[8] but merely observe that if his claim is true, that would mean that we can, in principle, eliminate the ontological commitments of applied mathematics to abstract objects!

---

[7] "I agree with Boolos [1985] and others, however, that we do have an independent grasp of plural quantifiers and their accompanying constructions, and that we can use them to formulate many truths at any given level that would have to be regarded as false on an ontology that repudiates classes of objects at that level." G. Hellman, *Structuralism without structures*, "Philosophia Mathematica" 1996, vol. 3(4), pp. 100–123.

[8] Cf. G. Hellman, *Toward a Modal-Structural Interpretation of Set Theory*, "Synthese" 1990, 84, pp. 409–443 for a detailed technical presentation.

4. Final Remarks

Similar problems emerge, when we consider other logics, e.g. logics with the branching quantifiers (e.g. Henkin quantifiers).[9] The translation of sentences with branching quantifiers into linear form involves quantification over functions. The problem arises, whether we should treat the linear form as **the genuine** logical form (and therefore treat sentences with the branching quantifier as – in a sense – deviant sentences), or rather consider branching quantifiers to be as legitimate as first-order ones. This problem is often discussed in the context of natural language, and the question to decide is whether quantification in natural language (and some other phenomena, as cross-references) is properly represented within standard, first-order logic.

Depending on our decision, our ontological commitments change, as it is not necessary to quantify functions if we accept branching quantifiers. The decision has – as in the case of the Boolos quantifier – an "ideological" character, depending on our acceptance of certain semantic notions as primitive.[10]

I think that these remarks clearly show that there is some interplay between ontology and ideology. A strengthening of the ideological assumptions makes it possible to decrease ontological commitments. I am convinced that this is a problem which deserves thorough discussion.

---

[9] For a discussion cf. e.g. J. Barwise, *On Branching Quantifiers in English*, "Journal of Philosophical Logic" 1979, vol. 8, pp. 47–80.

[10] A slightly different example is the case of modal notions, which are involved in certain nominalistic reconstructions of mathematical theories (e.g. Hellman's or Chihara's systems, cf. G. Hellman, *Mathematics without Numbers*, Clarendon Press, Oxford 1989; C. Chihara, *Constructibility and Mathematical Existence*, Clarendon Press, Oxford 1990). In this case we also accept certain modal notions as primitive, and not reducible to set-theoretic semantics.

Bartosz Brożek
Jagiellonian University
Copernicus Center for Interdisciplinary Studies

Neuroscience and Mathematics
From Inborn Skills to Cantor's Paradise[1]

Introduction: The Mathematical Subject

When one contemplates the history of the philosophy of mathematics, three kinds of mathematical subjects – i.e. subjects capable of carrying out mathematical reasoning – may be identified: the Platonic, the transcendental and the empirical.[2] The Platonic subject has the ability to 'see' mathematical objects, which belong to an independent, eternal reality. A 'strong' Platonic subject (God?) is capable of perceiving all the mathematical world at once, and so is capable of contemplating actual infinity; the 'weak' Platonic subject has some cognitive access to the mathematical universe – through some special faculty of mind such as intuition – but not all-encompassing.

The second kind of mathematical subject is the transcendental, a conception advocated vividly by Immanuel Kant and his followers (who include the most important philosophers of mathematics,

[1] The publication of this paper was made possible through the support of a grant "The Limits of Scientific Explanation" from the John Templeton Foundation. It is largely based on, and summarizes some parts of, the argument developed in B. Brożek, *Rule-following. From Imitation to the Normative Mind*, Copernicus Center Press, Kraków 2013.
[2] See A. Olszewski, *Teza Churcha. Kontekst historyczno-filozoficzny*, Universitas, Kraków 2009, chapter 6.2, *passim*.

such as Brouwer, and – with some reservations – even Hilbert). The transcendental subject *constructs* mathematical objects, and so – in a way – they are not independent of it. However, the use of the word 'construct' may be misleading here. The transcendental subject is not a physically existing – nor physically realizable – entity. It is an ideal projection of human mathematical capacities. In other words, the transcendental subject stands *vis a vis* the entire body of possible mathematical knowledge; it is a postulate of what *can be done* in mathematics, disregarding any physical limitations such as time or space. Thus, when we speak of 'constructions of mathematical objects', we are referring to what is constructable in principle, not actually. To put it in a different way: human mathematical practice not only can never transgress the boundaries set by the ideal of the transcendental subject, but it also can never reach them. Because of its characteristics, the transcendental subject cannot perceive (construct) actual infinity, but is capable of grasping potential infinity.

Finally, there exists the notion of an empirical mathematical subject, i.e. someone whose mathematical capacities are limited by spatio-temporal boundaries. For instance, a universal Turing machine is *not* a model of an empirical subject, as it utilizes an infinite tape and is capable of repeating its simple operations a number of times which is beyond the reach of any spatio-temporally limited agent.

It is my claim that – despite some attempts of the representatives of psychologism and similar stances – most of the philosophies of mathematics developed during the previous 200 years presuppose a Platonic or transcendental view of the mathematical subject. This is, naturally, an oversimplification, but since the rejection of psychologism at the turn of the 20th century, the abilities of mathematical subjects (e.g., Brouwer's creative subject, or the subject capable of manipulating symbols within Hilbert's formal systems) have always been understood in a non-empirical way. The spectacular advances in the neuroscientific studies of mathematical skills, achieved during the last two decades require, however, a re-thinking of the problem of the mathematical subject. So: is the empirical subject back?

In what follows I will sketch a conception of mathematics that is suggested by recent scientific findings. I will also identify some challenges to this view, and argue for a Popperian ontology of mathematics. I will conclude with an answer to the question posed at the end of the previous paragraph.

1. The Number Sense or the 'Embrained' Mathematics

During the last twenty years or so, a number of neuroscientific studies have been devoted to uncovering the origins of human mathematical capacities. The experiments in question include preferential looking, habituation of looking time, anticipatory head turning, explanatory reaching, neuroimaging with EEG or fMRI, but also the careful study of the mathematical skills of various animals (birds, non-human primates).[3]

To illustrate: an experiment of Izard *et al.* had the following setting: newborns were presented with a series of syllable trains of various pitch and duration. Half of the participants were exposed to 4-syllable trains, the other half – to 12-syllable trains. After a two minute time interval the infants were shown screens with visual arrays consisting of 4 or 12 objects. Their times of looking at the screen were recorded. It proved that they tended to look longer at the visual arrays that correspond to an ongoing sound sequence.[4]

Another example are the classical experiments of Starkey and Cooper. Infants were shown consecutive slides with two dots, but differing in size and the distance between them. The experimenters measured fixation time of the infants, i.e. the time they spent looking at a new slide. After a while the fixation times decreased – the new slides with two dots were looked at for shorter periods. Then, without

---

[3] Cf. E.S. Spelke, *Natural Number and Natural Geometry*, [in:] *Space, Time and Number in the Brain*, eds. S. Dehaene, E. Brannon, Academic Press, London 2011, pp. 287–317.

[4] *Ibid.*, p. 310.

a warning, slides with three dots appeared, and the infants exhibited longer fixation times, which suggested that they could perceive the difference between two and three.[5]

Yet another example of studying infants' mathematical abilities consisted in placing two puppets on a stage, covering them with a screen, and then visibly removing one of the puppets. When the screen was lowered and there was only one puppet, infants expressed no surprise; however, if there were still two puppets, the fixation time was longer, suggesting that the infants were surprised by what they saw.[6]

These and other results suggest that there exists an inborn capacity to 'deal' with small numbers; that this capacity is cross-modal (i.e., it 'extracts' numerosity independent of the mode of presentation – infants 'know' that three dots and three spoken syllables share the same numerosity); and that it enables understanding of simple mathematical operations performed on small numbers (addition or subtraction).

In fact, the current research implies that two separate brain systems are responsible for these capacities: the object tracking system (OTS) and the approximate number system (ANS). OTS is a system that enables tracking multiple individuals (up to 3 or 4). It is based on the principles of cohesion (moving objects are recognized as bounded wholes), continuity (objects move on unobstructed paths) and contact (objects do not interact at a distance).[7] The existence of the OTS system is confirmed by a number of tests, including visual short-term memory tasks, multiple-objects tracking tasks, or enumeration tasks. The last kind of tests confirms human ability of *subitizing*, i.e. of an instant and highly accurate determination of a number of object in small collections (3–4), even presented very briefly.[8] Furthermore,

---

[5] Cf. G. Lakoff, R. Núñez, *Where Mathematics Comes From*, Basic Books, New York 2000, p. 16.

[6] *Ibid.*

[7] M. Piazza, *Neurocognitive Start-Up Tools for Symbolic Number Representations*, [in:] *Space, Time and Number...*, op.cit., p. 270.

[8] *Ibid.*, p. 271.

it is speculated that the posterior parietal and occipital regions of the brain play the crucial role in the performance of such task, which suggests that these regions are the location of OTS.[9]

ANS, on the other hand, is a system for representing the approximate number of items in sets. It works according to the famous Weber's Law: the threshold of discrimination between two stimuli increases linearly with stimulus intensity. The law postulates a logarithmic relation between the physical stimulus and its internal representation. In the case of ANS, the Weber fraction, or the smallest variation to a quantity that can be readily perceived, changes over human development. For newborns it is 1:3, for 6-month-old babies it is 1:2, for 1-year-old children it is 2:3, for a 4-year-old it is 3:4, for a 7-year-old it is 4:5, while for a 20-year-old it is 7:8. It means that a newborn can discriminate between 1 and 3, or 2 and 6, or 10 and 30, but not 1 and 2, 2 and 5, or 10 and 27. Four-year-old children can tell that there is a difference in numerosity between sets consisting of 6 and 8 or 12 and 16 elements, but not 7 and 8 or 12 and 15. An adults' ANS system is even more 'sensitive': they can discriminate (without counting) between sets consisting of 14 and 16 elements or 70 and 80 elements, but not 70 and 78 elements. It is quite well established that ANS is located in the mid-intreparietal sulcus.[10]

One important difference between OTS and ANS must be noted here. ANS may be considered a core system *dedicated to numbers*. Its only task is to assess the approximate numerosities of sets. OTS, on the other hand, serves object tracking, and 'takes notice' of various aspects of the tracked objects such as their shape, colour, etc.

The existence of basic mathematical skills was also found in animals. Interestingly enough, their numerical skills are ratio-dependent, suggesting that they are equipped with a predecessor of the human ANS system. For instance, it has been demonstrated that "macaque monkeys spontaneously match the approximate number of individuals

9 *Ibid.*, p. 270.
10 *Ibid.*, pp. 268–269.

they see to the number of individual voices they hear; and also sum up visual and auditory stimuli to estimate their total number, without previous training."[11] This suggests that ANS is an old evolutionary device.

There are also other basic problems connected to mathematical thinking that generate much interest among neuroscientists. One such example is the problem of how the representation of numbers and complex counting is possible, given that working memory can store up to only 4 visual items. Feigenson speculates that this is possible thanks to the flexibility of the working memory – various things can count as items: individual objects (representations of exact numbers of objects), sets of objects or ensembles (representations of an approximate number of objects).[12] Another problem is the question of whether the process of counting depends on eye movement, as some suggest. Cavanagh and He claim that eye movement is not essential to counting; what is required is a set of attention pointers that individuate targets of interest in the visual field.[13]

Much attention is also devoted to the relationship between the representations of space, time and number. The research in this area was stimulated by the discovery of the SNARC effect by Dehaene and his team in 1993.[14] The experiment setting was as follows: adults were to decide whether the displayed number (in Arabic notation) is odd or even. In the former case, they were to press the left button, in the latter – the right one. The interesting observation was that large numbers led to faster key presses on the right-hand side of the space, irrespective of whether they were odd or even. This suggested that there is some

---

[11] *Ibid.*, p. 269.

[12] Cf. L. Feigenson, *Objects, Sets, and Ensembles*, [in:] *Space, Time and Number...*, *op.cit.*, pp. 13–22.

[13] Cf. P. Cavanagh, S. He, *Attention Mechanisms for Counting in Stabilized and in Dynamic Displays*, [in:] *Space, Time and Number...*, *op.cit.*, pp. 23–25.

[14] Cf. S. Dehaene *et al.*, *The Mental Representation of Parity and Number Magnitude*, "Journal of Experimental Psychology: General" 1993, vol. 122, pp. 371–396; W. Fias, J.-Ph. van Dijck, W. Gevers, *How Is Number Associated with Space? The Role of Working Memory*, [in:] *Space, Time and Number...*, *op.cit.*, pp. 133–148.

relationship between number and space representations. In particular, it may be speculated that "mental representation of numbers takes the form of a horizontally oriented line which is functionally homeomorphic to the way physical lines are represented"[15] (the so-called number line hypothesis). This would explain why large numbers lead to the preference of the right-hand side (on the number line, the larger the number is, the more to the right it is located). Other SNARC effects have also been discovered (associating numbers not only with space, but also with temporal duration, objects size or finger location) leading, e.g., to Vincent Walsh's ATOM theory (a theory of magnitude), which postulates a single magnitude mechanism that underlies the representations of number, space and time. Other researchers disagree with this proposal, suggesting other solutions, such as partially shared accumulator systems (Deheane), or the implication of the role of the serial position in working memory as an important determinant of the interactions between number and space (Fias).[16]

Altogether, the current research in the neurobiological foundations of mathematical knowledge clearly suggest that there exists an inborn – or, to coin a word, an *embrained* – set of very basic mathematical skills, which – in the context of arithmetic capabilities – consists of OTS and ANS. It must also be noted that similar inborn basic skills are implicated in the case of geometrical reasoning (the 2D and 3D systems).[17] For the sake of readability, I limit myself to the arithmetic capacities, but the conclusions I draw below are applicable, *mutatis mutandis*, to geometry.

---

[15] W. Fias, J.-Ph. van Dijck, W. Gevers, *How is Number Associated with Space? The Role…*, *op.cit.*, p. 133.

[16] See *ibid.*, pp. 139–144.

[17] Cf. E.S. Spelke, *Natural Number…*, *op.cit.*

2. The Bumpy Road to Cantor's Paradise
or Embodied Mathematics

The inborn or 'embrained' arithmetic is quite limited. The question thus is, how to get from the OTS and ANS systems to the beautiful but complex world of contemporary mathematics, with its differential equations, Noether rings, Hilbert spaces, non-commutative geometries, quaternions, and infinities. It is no surprise that – at least for now – neuroscience has little to say about this problem, and what it says is quite speculative.

Even the simplest of issues is controversial: what is the mechanism that enables us to break the 'number 4' barrier and acquire the competence to use natural numbers? Developmental psychology tells us that this process is very slow. At 6 months, children can add and subtract one; at two they begin to learn sequences of counting words, but do not map the words onto the numbers they represent; half a year later they recognize that number words mean more than one; at four, children can use fingers to aid adding; at five and a half they understand the commutativity of addition; a year later they understand the complementarity of addition and subtraction. More importantly, between 2. and 4. year, children learn to map number words '1', '2', '3' and '4' to the corresponding cardinalities one after another, but it may take up to six months to move to the next number.[18]

The first hypothesis of how an individual moves from counting from 1–4 to greater numbers has been developed by Carey. She calls the process bootstrapping. It consists in realizing that – when a children hears 'five' – the approximate numerosity of 'fiveness' is activated (in ANS) and then – on the basis of what the child knows about numbers 1–4 (from the workings of the 'exact', OTS system) – she concludes that 5 is also an exact number. Carey believes, then, that the essential move from '4' to '5' results from the interplay between both core mathematical systems: OTS and ANS. A child learns

---

[18] Cf. *ibid.*

what are exact numbers thanks to the way OTS operates; then, on the basis of the approximate representation of 'fiveness', a kind of inductive step is made, which enables the move to greater natural numbers.[19]

A different proposal has been developed by Piazza.[20] She observes that the ANS system may be used to represent not only large numbers, but also small ones. Moreover, ANS quite quickly becomes very precise as regards small numerosities. Given the progression in the sensitivity of ANS (the changes to the Weberian ratio), in order to distinguish between 2, 3, and larger numbers a ratio of 3:4 is needed. This happens at around three years of age, and coincides with the period when children become 'three-knowers'. In other words, Piazza believes that no interplay between OTS and ANS is needed to 'break the number four threshold' – the increasing precision of the ANS system is sufficient to account for this ability. She supports this hypothesis with the following observations. Firstly, it is true that before children learn how to count, their arithmetic abilities are limited to the initial four natural numbers; on the other hand, however, there is no evidence suggesting that OTS plays any role after counting skills are acquired. It is thus difficult to understand how a system that serves no purpose in full-blooded counting would be so essential in acquiring the counting skills. Secondly, the fact that it takes some time for a child to get from understanding that the word 'one' corresponds to the number 1, to understanding that the word 'two' corresponds to the number two, and then again from 2 to 3, and from 3 to 4, may serve as an argument against the hypothesis that OTS plays some significant role in the development of numerical skills – if it did, it would be natural to assume that children learn the association of number words with the four corresponding numbers at the same age. Thirdly, there is no data suggesting that dyscalculia, or the mental disorder resulting in problems with counting, has anything to do with the workings

---

[19] M. Piazza, *Neurocognitive Start-Up Tools…*, *op.cit.*, p. 276.

[20] *Ibid.*, pp. 275–276.

of OTS. Finally, experiments with infants and young children suggest that some features of OTS may be detrimental in numerical tasks. As pointed out above, OTS serves to track objects, 'taking notice' of their various characteristics (shape, size, colour, etc.): in some tasks, when such features (surface or contour length) were pitted against number, infants automatically attended to them and failed to recognize number. "It is extremely difficult to understand how a system that often interferes with numerical tasks might be relevant for learning yet more complex numerical representations."[21]

At the same time, it must be noted that the functioning of ANS alone is insufficient to account for the acquisition of counting skills. Even if Piazza is right, and the role of OTS is negligible in this process, ANS may only provide exact enough mechanism to learn the counting of relatively small numbers. As the Weberian ratio suggests, a 20-year-old person would be able to distinguish between 7 and 8, but not 8 and 9, so – if her counting abilities were based solely on ANS – she could use exact counting in the range of count from 1 to 8 only. This suggests that even if OTS plays no significant role in breaking the number 4 threshold on the top of ANS an additional mechanism is needed to explain the acquisition of natural number counting skills.

A hypothesis which addresses this problem is defended by Spelke. She observes that "children appear to overcome the limits of the core number system when they begin to use number words in natural language expressions and counting."[22] Children learn the first ten counting words by the age of 2, but initially use them without the intended meaning. At 3 they know that 'one' means one; at 4 they associate '2', '3' and '4' with the corresponding numerosities. Then, there is a kind of 'jump' – children learn quite quickly subsequent numbers. This, according to Spelke, requires two things: (a) to understand that every word in the counting list designates a set of individuals with a unique cardinal value; and (b) to grasp the idea that each cardinal value can

---

[21] *Ibid.*, p. 280.
[22] E.S. Spelke, *Natural Number…*, *op.cit.*, p. 304.

be constructed through progressive addition of 1.[23] How is this possible? "For most children, the language of number words and verbal counting appears to provide the critical system of symbols for combining the two core systems (i.e., ANS and OTS), and some evidence suggests that language may be necessary for this construction."[24]

Spelke's hypothesis that the development of counting skills is conditioned and mediated by the acquisition of language, is supported by the following pieces of evidence. First, both children and adults in remote cultures, whose languages have no words for numbers, recognize their equivalence only approximately when dealing with numbers larger than three. Second, an interesting line of evidence comes from the study of the mathematical abilities of deaf people. Deaf people living in numerate cultures but not exposed to the deaf community use a gestural system called homesign; they use their fingers to communicate numbers, but only with approximate accuracy. Similarly, they perform matching tasks with approximate accuracy.[25]

Spelke claims further that language continues to play an important role in mathematical cognition even after mathematical skills are mastered. For instance, educated adults who suffer language impairments have problems with exact, but not approximate numerical reasoning. Similarly, when doing exact (but not approximate!) tasks adults spend more time with numbers that are difficult to pronounce, even if they are presented in Arabic notation. But "if language merely scaffolded the acquisition of natural number concepts and abilities, and then was replaceable by other symbol systems, one would not expect adults to translate Arabic symbols into words for purposes of exact computation."[26] Finally, bilingual adults who are taught some new mathematical facts in one of their languages have difficulties in the smooth production of exact number facts in the other language.

---

[23] *Ibid.*, p. 305.
[24] *Ibid.*, p. 305.
[25] *Ibid.*, pp. 306–307.
[26] *Ibid.*, p. 307.

To be sure, there are serious doubts as to the extent of the role of language in the development of mathematical skills. Rochel German and Brian Butterworth believe that there is a "need to distinguish possession of the concept of numerosity itself (knowing that any set has a numerosity that can be determined by enumeration) from the possession of representations (in language) of particular numerosities."[27] Their case rests on the following observations. Firstly, they observe that the brain systems involved in numerical processing are 'some distance away' from the brain areas associated with the production of speech. Secondly, they are not convinced by the studies of Amazonian natives who have limited number vocabulary: their apparent lack of exact counting skills may result from the fact that they did not understand the tasks to be performed. In a latter study, Butterworth and colleagues compared arithmetic skills of three groups of children: English speaking from Melbourne, and two groups of indigenous Australians, speaking Warlpiri and Anindilyakwa. Warlpiri "has three generic types of number words: singular, dual plural, and greater than dual plural."[28] Anindilyakwa, in turn, "has four possible number categories: singular, dual, trial (which may in practice include four), and plural (more than three)."[29] The tests included: memory for a number of counters, cross-modal matching of discrete sounds and counters, nonverbal exact addition, and sharing play-dough disks that could be partitioned by the child. The results showed no 'language effects': all three groups performed similarly. This led Butterworth *et al.* to the conclusion that "the development of enumeration concepts does not depend on the possession of a number-word vocabulary. [Therefore] we are born with a capacity to represent exact numerosities, and (...) using words to name exact numerosities is useful

---

[27] R. Gelman, B. Butterworth, *Number and Language: How Are They Related?*, "TRENDS in Cognitive Sciences" 2005, vol. 9, no.1, p. 6.
[28] B. Butterworth, R. Reeve, F. Reynolds, D. Lloyd, *Numerical Thought with and without Words: Evidence from Indigenous Australian Children*, "Proceedings of the National Academy of Sciences" 2008, vol. 105, no. 35, (10.1073 PNAS 080604510), p. 13179.
[29] *Ibid.*, p. 13179.

but not necessary. When children learn to count, they are learning to map from their pre-existing concepts of exact numerosities onto the counting word sequence. Conceptual development drives the acquisition of counting words rather than the other way around."[30]

While Butterworth and his team's reservations seem well--founded, their conclusion seems to go too far. Firstly, the controversy does not concern the question of whether all of our mathematical abilities are inborn or else are enabled by language: we are speaking of relatively simple arithmetic tasks. It follows, secondly, that the thesis that language is crucial in the development of mathematical skills is not contested; rather, the extent of its role is at stake. Finally, Butterworth's team criticize the idea that all mathematical representations are conditioned on language representations or that – in the development of mathematical skills – language is *first* and *precedes* the emergence of mathematical mental concepts. However, this argument presupposes a certain view of language as based on an isolated system of mental representations. As I shall argue below, there exists a different approach to language, one regarding language as *not preceding* other forms of cultural representation, but rather *intertwined* with them. All in all, one should agree with Stanislas Dehaene who remarks:

> We all start out on the same rung, but we do not all climb to the same level. Progress on the conceptual scale of arithmetic depends on the mastery of a toolkit of mathematical inventions. The language of numerals is just one of the cultural tools that broaden the panoply of available cognitive strategies and allow us to resolve concrete problems.[31]

The claim that it is language and the other 'cultural tools' that shape our exact mathematical thinking is quite general. The question is,

---

[30] *Ibid.*, p. 13182.
[31] S. Deheane, *The Number Sense*, 2nd edition, Oxford University Press, Oxford 2011, pp. 263–264.

how exactly the passage from elementary mathematical operations on small numbers – *via* language – to transfinite numbers, axiomatic set theory or non-Euclidean geometries is made. There are not many proposals as how to answer this question. The most notable one is George Lakoff and Rafael Núñez's idea of *embodied mathematics* presented in their study *Where Does Mathematics Come From?*[32]

Lakoff and Núñez's work belongs to the approach known as the embodied mind paradigm. The basic tenets of this stance are as follows. The very idea of embodiment boils down to the thesis that the human mind and human cognition are decisively shaped by the experiences of our bodies. This is a vague claim that only underlines the rejection of other paradigms, such as Cartesian mind-body dualism or computationalism (the rough idea that the human brain is hardware, and the mind is software implemented in the brain). However, other claims of the representatives of the embodied mind approach are more informative. They believe that the human mind is a powerful conceptual system shaped by our bodies' experiences during their interactions with the environment. The most basic mental concepts or schema, probably derived from the neural motor-control programs, express spatial relations (such as the Source – Path – Goal schema). Since such "image schemas are conceptual in nature, they can form complex composites. For example, the word 'into' has a meaning – the *Into schema* – that is the composite of an *In schema* and *To schema.*"[33] Further, this mental machinery is capable of producing abstract concepts with the use of concrete ones through the use of metaphors. In the embodied paradigm metaphors are understood as the means for "understanding and experiencing one kind of thing in terms of another."[34] And so, importance is conceptualized in terms of size ("This is a big issue", "It's a small issue; we can ignore it"), dif-

---

[32] Cf. G. Lakoff, R. Núñez, *Where Mathematics Comes From?*, Basic Books, New York 2000.
[33] *Ibid.*, p. 39.
[34] *Ibid.*, p. 5.

ficulties are conceptualized as burdens ("He is overburdened", "I've got a light load this semester"), etc.[35]

> Each such conceptual metaphor has the same structure. Each is a unidirectional mapping from entities in one conceptual domain to corresponding entities in another conceptual domain. As such, conceptual metaphors are part of our system of thought. Their primary function is to allow us to reason about relatively abstract domains using the inferential structure of relatively concrete domains.[36]

Lakoff and Núñez claim further that it is the mechanism of conceptual metaphorization that enables the construction of complex and precise mathematical concepts. In the case of arithmetic, they postulate the existence of four basic or grounding metaphors: the Arithmetic as Object Collection (where the source domain concept of collections of objects of the same size is mapped to the concept of numbers, the size of the collection is mapped to the size of the number, the smallest collection is mapped to the concept of the unit, while putting collections together is mapped to the process of addition); the Arithmetic as Object Construction (where the source domain concept of objects consisting of ultimate parts of unit size is mapped to the concept of numbers or the act of object construction is mapped to the concept of arithmetic operations); the Measuring Stick metaphor (where physical segments are understood as numbers, the basic physical segment as one, and the length of the physical segment as the size of the number); and the Arithmetic as Motion Along a Path metaphor (where the act of moving along the path is understood as representing mathematical operations, point-locations on the path are understood as numbers, etc.).

Lakoff and Núñez claim that those four grounding metaphors give rise to the development of basic arithmetic. One begins with innate capacities to 'deal' with small numbers (up to 4). In addition, one

---

[35] *Ibid.*, p. 41.
[36] *Ibid.*, p. 42.

has primary experiences with object collections, object construction, physical segmentation and moving along a path. "In functioning in the world, each of those primary experiences is correlated with subitizing, innate arithmetic, and simple counting."[37] Those two domains are combined through the four metaphors in such a way that the primary experiences become sources of the metaphors and the domain of numbers is the target of the metaphors.

> The next step is the conflation among the primary experiences: object construction always involves object collection. Placing physical segments end to end is a form of object construction (…). From a neural perspective, [such conflations] involve co-activations of those brain areas that characterize each of the experiences. (…) As a consequence, an isomorphic structure emerges across the source domains (…), which is independent of numbers themselves and lends stability to arithmetic.[38]

The ability of subitizing – found in all normal human beings – leads to precise and stable results regarding small numbers; when extended with the four grounding metaphors, the precision and stability extends to all natural numbers. Finally, "the laws of arithmetic (commutativity, associativity and distributivity) emerge first as properties of the four source domains, then as properties of numbers via those metaphors, since the metaphors are inference-preserving conceptual mappings."[39]

Lakoff and Núñez use the same ideas to explain the emergence of more complex mathematical concepts, such as algebra, logic and set theory, real numbers, etc. There is no need to present here all those constructions, as they remain highly speculative. It is worthwhile, however, to have a look at one example: actual infinity.

---

[37] *Ibid.*, p. 93.
[38] *Ibid.*, pp. 95–96.
[39] *Ibid.*, p. 96.

We hypothesize – Lakoff and Núñez say – that all cases of actual infinity – infinite sets, points at infinity, limits of infinite series, infinite intersections, least upper bounds – are special cases of a single conceptual metaphor in which processes that go on indefinitely are conceptualized as having and end and an ultimate result. We call this metaphor the *Basic Metaphor of Infinity*. The target domain of the BMI is the domain of processes without end – that is, what linguists call imperfective processes. The effect of BMI is to add a metaphorical completion to the ongoing process so that it is seen as having a result – an infinite *thing*.[40]

In the case of BMI, the source domain are the iterative processes in which, although they do have a finite number of iterations, the number is indeterminate. The essential element of the metaphor is, however, to "add to the target domain the completion of the process and its resulting state."[41]

The theory developed by Lakoff and Núñez may be criticized in a number of ways. One can point out some mathematical errors in the book (e.g., the 'invention' of 'the first infinitesimal'). It may also be pointed out that Lakoff and Núñez seem to underestimate the role of ANS in arithmetic skills acquisition, while overplaying the role of OTS. One may even try to dismiss their approach claiming that these are 'just so stories', without any solid empirical grounds with the exception of some linguistic insights (e.g., the fact that we speak of zero using words which are connected to emptiness, lack, absence or destruction, and that number one has connotations with individuality, separateness, wholeness, integrity or beginning).[42] Further, Lakoff and Núñez's idea of schemes and concepts as constituting the 'deep structure of language' may be to some extent mistaken. However, all this misses the crucial point. Unlike other theorists, Lakoff and Núñez

---

[40] *Ibid.*, p. 158.
[41] *Ibid.*, p. 158.
[42] *Ibid.*, p. 75.

do offer an account of how to get from the limited innate mathematical capacities to full-blooded mathematics (even if they do not describe this process in the flawless way). Moreover, they show how it is possible to get to Cantor's paradise by starting with more *concrete* concepts, which reflects both phylogenetic and ontogenetic trajectories described in the literature.

All in all, Lakoff and Núñez's conception is a captivating idea and, in particular, their conception of *embodiment*: mathematical cognition is embodied in the sense that "it is grounded in simulations of sensorimotor processes through the use of neural resources that are also active in bodily perception and action." Embodied mathematics – in contrast to more modest approaches – does draw a line from the workings of OTS and ANS to our mathematical practices. It is a road that goes through our concrete bodily experiences as reflected in language; in a way, then, one can think of their theory as a concretization of Spelke's bold claim that the road to Cantor's paradise leads through language.

3. On How to Remain in Paradise
or the Embedded Mathematics

The best way to assess the embodied approach to mathematics is to compare it with an alternative. Given that there seems to exist a strong link between the acquisition of language and the development of mathematical skills, it is advisable to see how the origins of mathematical concepts may be explained within different approaches to the evolution of language.

Let us begin with the Chomskyan idea of universal grammar. As well known, Chomsky claims that some basic grammatical structures are hard-wired in the brain: all languages must have a common structural underpinning. The question is whether the Chomskyan approach may be helpful in explaining where mathematics comes from, or, in other words, how to get from our inborn mathematical capacities (ANS and OTS) to our actual mathematical practices. It seems to me that the answer is a plain 'no'. There is no route from simple mathematical capacities to full-blooded mathematics through the assumption that we have some hardwired grammatical rules. Moreover, I believe that this argument can be generalized: any conception which posits some formal inborn structures (proto-language, universal grammar, proto-language) is incapable of paving the way from the embrained mathematics to Cantor's paradise. First, it seems that one cannot take advantage of such inborn formal capacities to break the 'number four threshold" and, in consequence, to understand how counting to natural numbers greater than 4 is possible. One may object here, claiming that if the proto-logic or proto-language has a built-in induction mechanism, it would explain the move from number 4 to larger numbers. However, the problem with breaking the 'number four threshold' has little to do with induction; the problem is rather how to grasp the idea that numbers larger than 4, represented approximately by the ANS system, are exact numbers in the same way numbers 1–4 are. Thus, it is rather a 'material' or 'substantive' issue than a formal one. Further, even if we assumed that some inborn universal formal structures may explain the acquisition of the arithmetic skills, the same cannot hold of geometry. It seems, therefore, that Chomsky-like theories of language cannot help us in accounting for the development of mathematical thinking.

Fortunately, there exists an alternative explanation of language evolution and acquisition. It is advocated by such scholars as Merlin Donald, Michael Tomasello, Michael Arbib and others.[43] The fundamental observation motivating this stance is that the genetic difference between the human species and other animals is not so large (appox. $1 - 1.2\%$). This constitutes an argument for the thesis that the

---

[43] Cf. M. Tomasello, *Cultural Origins of Human Cognition*, Harvard University Press, Cambridge, Mass. 2001; M. Tomasello, *Why We Cooperate*, MIT Press, Cambridge, Mass. 2009; M. Arbib, *The Mirror System, Imitation, and the Evolution of Language*, [in:] *Imitation in Animals and Artifacts*, eds. Ch. Nehavin, K. Dautenhahn, MIT Press, Cambridge, Mass. 2002, pp. 229–280; M. Donald, *Imitation and Mimesis*, [in:] *Perspectives on Imitation*, vol. 2: *Imitation, Human Development, and Culture*, eds. S. Hurley, N. Chater, MIT Press, Cambridge, Mass. 2005, pp. 283–300.

biological adaptation enabling the flourishing of human culture must be relatively 'small'. In other words, the proponents of the described scenario claim that it is impossible to account for the development of various aspects of culture (language, morality, science, etc.) by recourse to a large number of biological adaptations. In particular, Michael Tomasello believes that:

> the 6 million years that separates human beings from other great apes is a very short time evolutionary, with modern humans and chimpanzees sharing something on the order of 99 percent of their genetic material – the same degree of relatedness as that of other sister genera such as lions and tigers, horses and zebras, and rats and mice. Our problem is thus one of time. The fact is, there simply has not been enough time for normal processes of biological evolution involving genetic variation and natural selection to have created, one by one, each of the cognitive skills necessary for modern humans to invent and maintain complex tool-use industries and technologies, complex forms of symbolic communication and representation, and complex social organizations and institutions.[44]

Tomasello further claims that

> there is only one possible solution to this puzzle. That is, there is only one known biological mechanism that could bring about these kinds of changes in behavior and cognition in so short a time (…). This biological mechanism is social or cultural transmission, which works on time scales many orders of magnitude faster than those of organic evolution.[45]

This, in turn, is made possible by three forms of learning: imitative, instructed and collaborative, which are conditioned by "a single spe-

---

[44] M. Tomasello, *Cultural Origins…*, *op.cit.*, p. 2.
[45] *Ibid.*, p. 4.

cial form of cognition, namely, the ability of individual organisms to understand conspecifics as being *like themselves*, who have intentional and mental lives like their own."[46]

There are a number of facts that underline the distinctive human capacity to relate to conspecifics. The strongest such line of argumentation is connected to the differences between humans and other primates. In a number of experiments, Tomasello and his colleagues have demonstrated that the great apes' learning differs substantially from human ways of cultural transmission. In particular, apes learn by emulation (i.e., they grasp only the means-ends structure of an activity and do not copy the pattern of behaviour), while humans learn by imitation and instruction.

These observations suggest that the *ability to imitate* is one of the crucial adaptations in the evolutionary history of humankind. Tied with this is the capacity of '*mindreading*' or '*intention-reading*'. In addition, evolution has equipped us in a cluster of emotions (for Tomasello, guilt and shame are the basic emotions for cementing social bonds). These adaptations, taken together, are responsible for what Tomasello calls human *mutualism*.[47] We not only have the ability to take the perspectives of others; we can also take a perspective *with* others. To put it differently, humans not only understand what some other individuals do (their intentions), but also do things *together* with them (i.e., we are capable of *we-intentionality*). This, according to Tomasello, is the key for understanding the possibility of cumulative cultural evolution.

A similar evolutionary scenario was sketched by Merlin Donald in relation to the emergence of language. Donald claims that the sources of the human ability to use language are based on *mimetic skills*, which evolved some 2 million years ago. He distinguishes between mimicry, imitation and mimesis. Mimicry is a simple copying of some action, with no understanding of its goal. Imitation is more

---

[46] *Ibid.*, p. 5.
[47] Cf. M. Tomasello, *Why We Cooperate, op.cit.*

abstract and flexible, as it takes into account the goal of the action. Finally, mimesis is defined as:

> the reduplication of an event for communicative purposes. Mimesis requires that the audience be taken into account. It also demands taking a third-person perspective on the actor's own behaviour. Some examples are children fantasy play, the iconic gestures used in a social context, and the simulation of a 'heroic' death during a theatrical performance.[48]

Mimetic skills are thus founded on the ability to imitate, which in turn is conditioned by the mimicry skills.

Donald identifies four main types of mimetic representation, which are key to the transmission and propagation of culture: (1) reenactive mime, characteristic of role-playing; (2) precise means--end imitation (as in learning how to fry an egg); (3) the systematic rehearsal and refinement of skill (as in learning how to drive a car); and (4) nonlinguistic gesture (as in learning how to dance). He claims further that these mimetic skills were the foundation for the emergence of language and all the other forms of culture. He stresses that his proposal differs from the traditional scenarios which condition the emergence of culture on the prior emergence of language (the *language first* theory). According to Donald, some forms of culture, based on the mimetic skills, must have *preceded* language and enabled its evolution (the *culture first* theory).

Donald's theory leads to profound consequences. First, he claims that the human mind is intimately linked to the society in which it flourishes. One can even say that it is *co-created* by the community. The communal practices are constitutive of the human mind, both in the phylogenetic and ontogenetic dimensions. Second, language is not an individual but a network-level phenomenon: its evolution resembles the evolution of an ecosystem rather than of a single organ-

---

[48] M. Donald, *Imitation and Mimesis, op.cit.*, p. 289.

ism. Third, it follows that "cognitive neuroscientists are unlikely to find an innate language acquisition device, and should redirect their investigations toward the powerful analogue processing systems out of which language can emerge in group interactions."[49]

Another analogous theory – but in the context of neuroscience – is advocated by Michael Arbib. He attempts to answer the question of what is the neuronal basis for a certain feature of language – *parity* – which manifests itself in our ability to recognize what our interlocutor wants to say.[50] He observes that Broca's area – traditionally considered as the region of the brain responsible for the production of speech – is one of the areas in which there is a complex system of mirror neurons. Thus, Broca's area is implicated in the production of various multimodal linguistic actions (utilizing the hands, face and voice). In connection to this, Arbib formulates the Neuron System Hypothesis: the parity condition is fulfilled due to the fact that Broca's area has been evolutionarily built upon a perception system responsible for the recognition and execution of manual actions. Arbib believes that the hypothesis is backed by both the arguments resulting from neuroimaging experiments (execution and perception of manual gestures activate neurons located within or in the proximity of Broca's area) and the anatomical facts (it is assumed that the F5 region in the macaques brains, where the mirror neurons were discovered, is an analogue of the Broadmann 44 area in human brain, which is a part of Broca's area).

Arbib also presents an evolutionary hypothesis pertaining to the probable development of mirror neurons. The first stage consisted in the emergence of mirror systems dedicated to the perception and execution of manual actions. In the second stage, those mirror neurons served as the basis for the development of the ability to imitate manual gestures: simple forms of such imitation are found in apes, more complex forms are exclusively human. The third stage was the emergence

---

[49] *Ibid.*, p. 294.
[50] Cf. M. Arbib, *The Mirror System…*, *op.cit.*

of pantomimic skills, and the fourth stage led beyond the simple reenactment of human behaviour (some gestures were recognized as standing for something else, e.g. waving one's hand was interpreted as imitating a flying bird). The fifth stage was the emergence of proto--signs, or conventional gestures which made pantomime more precise (e.g., they enabled to distinguish gestures representing birds and the process of flying). Finally, the sixth stage resulted in the development of proto-language, which emerged through the separation of conventionalized manual, mimic and vocal gestures from pantomime.[51]

Arbib claims further that if the above sketched evolutionary scenario is sound, the human brain is *language-ready*, but *does not 'have' language*. Thus, we are forced to reject Chomsky's theory of universal grammar which is a conception of a brain 'having language'. This conclusion is further reinforced by the following two observations: first, the ability to learn and use language is not confined to the spoken language, but embraces a combination of manual, vocal and mimetic skills.[52] Second, proto-language was based on proto-phrases: proto-words functioned as our sentences, not our words.

The major objection against this view of language – at least from the Chomskyan perspective – is that it may have problems dealing with the poverty of stimulus argument. Let me recall that Chomsky suggested his theory of universal grammar partly in response to the observation that children – in their development – are not exposed to a sufficient number of external stimuli in order to form – solely on this basis – a formal grammatical structure; thus, the argument runs, at least the most fundamental aspects of the structure must be innate.

The reply of the proponents of the imitation-based account of language rests on two theses. On the one hand, they claim that Chomsky failed to realize that spoken language is not an isolated system: it is a part of, and is based upon, a larger cluster of communication skills,

---

[51] See *ibid*.

[52] M. Arbib, *Mirror Neurons and Language*, [in:] *Handbook of the Neuroscience of Language*, eds. B. Stemmer, H.A. Withaker, Academic Press, London 2006, p. 238.

embracing, in particular, the language of gestures. Thus, the stimuli that need to be taken into account are not limited to vocal ones; to the contrary, communication vocal stimuli form just a small part of what a child is exposed to in the process of communication.

On the other hand, the mechanism of imitation is a powerful tool that – in certain circumstances – may lead to a kind of 'combinatorial explosion'. Giacomo Rizolatti claims that there are two types of mirror neuron resonance and, as a result, two types of imitation. The high-level resonance is used to mirror the *goal* of an action, while the low-level resonance copies the *way of acting*. According to Rizolatti, only the human brain takes advantage of both mechanisms, which enables imitation *sensu stricto*. In apes and (possibly) other animals only high-level resonance is used and this explains apes' learning by emulation only. The utilization of both high-level and low-level resonance plays a key role in the enhancement of the flexibility and stability of human social reaction, as it enables – through recombination – to use the same patterns of behaviour, learned by imitation, to realize different goals, or to realize one goal with different means.[53]

This theory may also explain the mechanism of 'novelty' in mathematics, or – if I may call it so – the Meno problem. The possibility of discoveries in mathematics are explicable by recourse to two facts. First, in order to develop new mathematical concepts, one utilizes the mechanism of metaphorization: a new metaphor may lead to the establishment of new mathematical facts. Second, the fact that we learn much of our mathematics through imitation suggests that through the combination of previously learned patterns of behaviour, new connections between mathematical means and ends may be established.

Interestingly, there are studies which suggest some close links between the conception of language and culture as originating with imitation and the 'embodied' paradigm as advocated by Lakoff and

---

[53] Cf. G. Rizolatti, *The Mirror Neuron System and Imitation*, [in:] *Perspectives on Imitation vol. 1: Mechanisms of Imitation and Imitation in Animals*, eds. S. Hurley, N. Chater, MIT Press, Cambridge, Mass. 2005, pp. 55–76.

Núñez. Lakoff and Gallese suggest[54] that "that the sensory-motor system has the right kind of structure to characterize both sensory-motor and more abstract concepts."[55] Their argument runs as follows. First, they observe that simple action concepts such as 'grasp' may well be characterized at the level of sensory-motor structure. Second, they claim that mirror neurons and other premotor and parietal neurons are 'multimodal', i.e. they respond to more than one modality. "Such multimodality meets the condition that an action-concept must fit both the performance and perception of the action."[56] Third, "multimodality is realized in the brain through *functional clusters*, that is, among others, parallel parietal-premotor networks. These functional clusters form *high-level units*—characterizing the discreteness, high-level structure, and internal relational structure required by concepts."[57] Fourth, they argue that in order to understand a concept one needs to imagine oneself or someone else doing what the concept refers to. They claim further that imagination is a mental simulation, "carried out by the same functional clusters used in acting and perceiving."[58] Therefore, any conceptualization of a concrete concept (e.g., grasping) *via* simulation requires the use of the same functional clusters used in the action and perception of grasping. Further,

> all actions, perceptions, and simulations make use of neural parameters and their values. For example, the action of *reaching* for an object makes use of the neural parameter of direction; the action of *grasping* an object makes use of the neural parameter of force. So do the concepts of *reaching* and *grasping*. Such neural parametrization is pervasive and imposes a *hierarchical structure* on the brain: the same parameter values that characterize the *internal*

---

[54] Cf. V. Gallese, G. Lakoff, *The Brain's Concepts: The Role of the Sensory-Motor System in Conceptual Knowledge*, "Cognitive Neuropsychology" 2005, vol. 22 (3–4), pp. 455–479.
[55] *Ibid.*, p. 455.
[56] *Ibid.*, p. 458.
[57] *Ibid.*, p. 458.
[58] *Ibid.*, p. 458.

structure of actions and simulations of actions also characterize the *internal structure* of action concepts.[59]

The final step of the argument is to move from action and other concrete concepts to more abstract ones. In this context, Lakoff and Gallese recall Narayanan's conception of certain premotor structures called X-schemas, which fit the perceptual structure of the motor actions. Moreover, Narayanan observed that the X-schemas "have exactly the right structure to characterize the collection of concepts that linguists refer to as 'aspect' – concepts that characterize the structure of events and our reasoning about events."[60] In this way we found ourselves in a known territory – recall that the X-schemas are fundamental to Lakoff and Núñez's theory of embodied mathematics. In other words, it may be suggested that some version of the imitation-based paradigm in the study of language constitutes the neural and evolutionary complement of Lakoff and Núñez's conception of the embodied mind and embodied mathematics.

It should be added that there is a considerable body of evidence linking bodily processes and mathematical cognition. First, "studies on neural correlates of hand movements and action understanding of hand gestures point to an overlapping circuitry in the prefrontal and intraparietal regions with number processing."[61] Second, "studies conducted with repetitive Transcranial Magnetic Stimulation (rTMS) show excitability of hand muscles during different number processing tasks."[62] Third:

> behavioral studies on math learning provide evidence for better math learning when instruction is supported with hand gestures;

---

[59] *Ibid.*, p. 458.
[60] *Ibid.*, p. 470.
[61] F. Soylu, *Mathematical Cognition as Embodied Simulation*, [in:] *Proceedings of the 33rd Annual Conference of the Cognitive Science Society*, eds. L. Carlson, C. Hölscher, T. Shipley, Cognitive Science Society, Austin, TX 2011, p. 1213.
[62] *Ibid.*, p. 1213.

higher problem solving performance when non-communicative hand gestures are allowed, compared to when hands are restricted; and non-communicative hand gestures during problem solving provide clues for misconceptions in conceptual understanding of arithmetic and algebra.[63]

The picture resulting from those observations is the following. The road from inborn mathematics to full-blooded mathematics leads through language. But language itself is not based solely on inborn grammatical structures. It is rather a part of a larger whole of communication and culture-generating skills which take advantage of the mechanism of imitation. If so, language is deeply rooted, or co-constituted by our social practices. In other words, in addition to being embodied (based on our bodily experiences), it is also *embedded* – it crucially depends on our communicating with others. If this is true of language, it is also true of our mathematical practices: they are embrained, embodied *and* embedded. Social interactions are not only *triggers* of mathematical concept-construction – they are *constitutive* of our mathematical knowledge.

It must also be stressed that the fact that our mathematics is embedded within social practices helps to explain why our mathematical practice is highly stable. Lakoff and Núñez claim that "the stability of embodied mathematics is a consequence of the fact that normal human beings all share the same relevant aspects of brain and body structure and the same relevant relations to their environment that enter into mathematics."[64] So, their claim is that, because our conceptual metaphors have the same source domain (our bodily experiences represented in the concrete concepts we use), or that the source domain is in a way *shared,* the resulting abstract mathematical concepts must also be shared, which explains why mathematics is so

---

[63] *Ibid.*, p. 1213.
[64] G. Lakoff, R. Núñez, *Where Mathematics Comes From*, Basic Books, New York 2000, p. 352.

stable. There is a defect in this argument: the mechanism of meta-phorization is such that it enables the development of numerous abstract concepts on the basis of the same concrete concept (as Lakoff and Núñez's examples clearly show). The question is, then, why people, when metaphorizing mathematics, pick the same (or similar) metaphors. The answer to this question is provided, I believe, by the theory of culture based on imitation. The fact that we learn cultural patterns of behaviour (and so also counting, theory-proving, etc.), contributes to the stability of our mathematical conceptual scheme: we use the same abstract metaphors, because we teach each other those metaphors, and we have an inborn capacity to learn in this imitative way and to propagate the learned patterns of behaviour among the members of our society.

There are also two philosophical corollaries of the theory sketched above. The first is that even if the road to Cantor's paradise is a bumpy one, it resembles more of a highway than a mountain route. Mathematics is not a stand-alone, separate body of knowledge. It is intimately intertwined with all that we call culture. It would be hopeless to try to distil the phylogenetic or ontogenetic path of the development of mathematical skills while not considering it an aspect of, or a 'fibre' in, the development of culture in general.

Secondly, the above presented theory – a highly speculative one, I admit – represents an interesting example of the use of the criterion of coherence within cognitive science. Neuroimaging data, experiments in developmental psychology, linguistic facts, evolutionary scenarios, etc – while taken in isolation – may usually be interpreted in a number of ways. However, when put together, they may strengthen or reinforce one another. This is, I submit, the case with the theory outlined here, in which the findings of neuroscience, linguistics and evolutionary theory contribute to a coherent picture of the origins of mathematical thinking.

4. Trimming Plato's Beard?

In the next two sections of the essay I will attempt to address two challenges that may be raised against the theory of embrained, embodied
and embedded mathematics, or – as I like to call it – the 3E theory.
These are: the problem of the necessity in mathematics, and the problem of the mathematicity of the universe.[65]

There is a dimension of mathematical and logical research that
traditionally poses a challenge to any naturalistic accounts of the ontology of mathematical or logical objects. It is well captured in the
following observation by Jan Łukasiewicz:

> Whenever I deal with the smallest logical problems, I always have
> the feeling that I am facing some powerful, incredibly coherent and
> enormously resistant structure. I cannot make any changes within
> it, I create nothing, but working hard I uncover new details, gain-
> ing eternal truths.[66]

Such views as the one expressed by Łukasiewicz above give rise to
the development of mathematical Platonism (or realism), a view that
"mathematics is the scientific study of objectively existing mathematical entities just as physics is the study of physical entities. The statements of mathematics are true or false depending on the properties
of those entities, independent of our ability, or lack thereof, to determine which."[67]

---

[65] There is one more challenge to the 3E account which I do not discuss here: the
'miraculous abilities' of mathematical prodigies (including *idiot savants*) and mathematical geniuses. This is a complex issue, but I believe that the challenges it poses
will ultimately be resolved through neuroscientific investigations. See especially, e.g.,
S. Deheane, *The Number Sense*, 2nd edition, Oxford University Press, Oxford 2011,
pp. 129–159.

[66] J. Łukasiewicz, *W obronie logistyki. Myśl katolicka wobec logiki współczesnej*, "Stu-
dia Gnesnensia" 1937, vol. 15.

[67] P. Maddy, *Realism in Mathematics*, Oxford University Press, Oxford 1990, p. 21.

There are many forms of mathematical Platonism. In particular, one should distinguish between ontological Platonism (a view pertaining to the existence of mathematical objects) and semantic Platonism (an epistemological view that mathematical statements are true or false). Ontological Platonism is a stronger theory – it implies the semantic one, but the opposite implication does not hold. Thus, in what follows I shall concentrate on the stronger claim. Arguably, ontological Platonism in mathematics, although it comes in various incarnations, embraces the following three theses:

(The existence thesis) Mathematical objects (or structures) exist.
(The abstractness thesis) Mathematical objects are abstract, non--spatio-temporal entities.
(The independence thesis) Mathematical objects are independent of any rational or irrational activities of the human mind. In particular, mathematical objects are not our constructions.[68]

The key question is how the above formulated theses are justified. With no pretence to comprehensiveness, I posit that there are three kinds of arguments backing mathematical Platonism in its ontological version. The first one is the *semantic argument*, well captured by Balaguer, but formulated earlier by Frege:[69]

(1) Mathematical sentences are true.
(2) Mathematical sentences should be taken at their face value. In other words, there is no reason to believe that mathematical sentences, as they appear, are not what they really are, or that there is a deep structure of mathematical sentences which differs from their surface structure, of what they seem at their face.

---

[68] Cf. Ø. Linnebo, *Platonism in the Philosophy of Mathematics*, [in:] *The Stanford Encyclopedia of Philosophy* (Fall 2011 Edition), ed. E.N. Zalta, 2011, URL = <http://plato.stanford.edu/archives/fall2011/entries/platonism-mathematics/>.
[69] Cf. *ibid.*

(3) By Quine's criterion, we are ontologically committed to the existence of objects which are values of the variables in the sentences we consider true.

(4) We are ontologically committed to the existence of mathematical objects.

(5) Therefore, there are such things as mathematical objects, and our theories provide true descriptions of these things. In other words, mathematical Platonism is true.

The second argument defending mathematical Platonism is *the indispensability argument*, or the Quine/Putnam argument. Maddy summarizes it: "we are committed to the existence of mathematical objects because they are indispensable to our best theory of the world and we accept that theory."[70] And in Putnam's own words: "mathematics and physics are integrated in such a way that it is not possible to be a realist with respect to physical theory and a nominalist with respect to mathematical theory."[71] A reconstruction of this argument might appear as follows:

(1) By Quine's criterion, we are committed to the existence of objects which our best physical theories speak of.

(2) Our best physical theories are expressed with the use of the language of mathematics.

(3) Therefore, we are committed to the existence of mathematical objects.

(4) When one is a realist with respect to physical theories, one must also be a realist with respect to mathematics.

(5) Therefore, mathematical Platonism is true.

Finally, Gödel's *intuition-based argument* may be reconstructed in the following way:

---

[70] P. Maddy, *op.cit.*, p. 30.
[71] H. Putnam, *What Is Mathematical Truth?* (1975), reprinted in H. Putnam, *Mathematics, Matter and Method*, Cambridge University Press, Cambridge 1979, p. 74.

(1) The most elementary axioms of set theory are obvious; as Gödel puts it, they "force themselves upon us as being true."[72]

(2) In order to explain (1), one needs to posit the existence of mathematical intuition, a faculty analogous to the sense of perception in the physical sciences.

(3) Not all mathematical objects are intuitable; but our belief in the 'unobservable mathematical facts' is justified by the consequences they bring in the sphere controllable by intuition and through their connections to already established mathematical truths. As Gödels says, "even disregarding the [intuitiveness] of some new axiom, and even in case it has no [intuitiveness] at all, a probable decision about its truth is possible also in another way, namely, inductively by studying its 'success' (…). There might exist axioms so abundant in their verifiable consequences, shedding so much light upon a whole field, and yielding such powerful methods for solving problems (...) that, no matter whether or not they are [intuitive], they would have to be accepted at least in the same sense as any well-established physical theory."[73]

Let us consider now whether the 3E account of mathematics defended in this essay has any bearing on the arguments favouring mathematical Platonism. Lakoff and Núñez believe that the conception of the embodied mathematics puts mathematical Platonism to eternal rest. For them, mathematical Platonism is "the romance of mathematics", a "story that many people *want* to be true"; a story that mathematical objects are real, and mathematical truth is universal, absolute, and certain. They succinctly reject this view:

---

[72] K. Gödel, *What Is Cantor's Continuum Problem?* (1947), reprinted in *Philosophy of Mathematics*, eds. P. Benacerraf, H. Putnam, Cambridge University Press, Cambridge 1983, p. 484.
[73] *Ibid.*, p. 477.

The only access that human beings have to any mathematics at all, either transcendent or otherwise, is through concepts in our minds that are shaped by our bodies and brains and realized physically in our neural systems. For human beings – or any other embodied beings – mathematics *is* embodied mathematics. The only mathematics we can know is mathematics that our bodies and brains allow us to know. For this reason, the *theory of embodied mathematics* (…) is anything but innocuous. As a theory of the only mathematics we know or can know, it is a theory of what mathematics *is* – what it really is![74]

As I read them, Lakoff and Núñez underscore two things. First, they put forward an epistemological claim that we have no cognitive access to independent abstract objects, since the only way of practicing mathematics is through the concepts "shaped by our bodies and brains." The problem of the cognition of abstract objects has been a subject of controversy since the beginnings of philosophy. Painting with a broad brush, one may claim that two solutions have been defended in this context, both already present in Plato: that there exists a rational intuition enabling us to contemplate abstract objects or that our access to the abstract sphere is discursive, mediated by language. Lakoff and Núñez seem to consider only the first option, and dismiss it on the basis of recent findings in neuroscience.

Second, they seem to embrace a version of Quine's criterion: we are committed to the existence of those things only, of which our best scientific theories speak. They add that the best – or rather: the only – theory of mathematical cognition we have is the theory of embodied mathematics, and since it does not speak of independent abstract objects, we have no grounds for postulating their existence. The problem is that Quine's criterion – applied to other theories, not necessarily accounting for the nature of mathematics, e.g. our best physical

---

[74] G. Lakoff, R. Núñez, *Where Mathematics Comes From*, *op.cit.*, p. 346.

theories – brings a different outcome: that we are indeed committed to the existence of abstract mathematical objects.

To put it differently: it seems that Lakoff and Núñez's stance does not invalidate any of the three arguments in favour of mathematical Platonism described above. In order to defeat the semantic argument, Lakoff and Núñez would either have to show that mathematical statements cannot be ascribed truth or falsehood (which they do not do); or to reject the idea that mathematical statements have no 'deep structure', that they are what they seem at their face (which they do not do as well); or to reject Quine's criterion of ontological commitment (which they *embrace* in their argument). Given the main idea behind the embodied approach, i.e. that mathematical concepts are the outcomes of metaphorization, their 'best shot' would probably be the rejection of the second claim (that there is no deep structure to mathematical statements). However, the distinction between the surface and deep structures of statements hangs together with some conception of the *form* of those expressions. It is the *form* of an expression that constitutes its deep structure. Thus, were the distinction to survive at all, Lakoff and Núñez would need to introduce a very peculiar conception of the form of statements which lies *beneath* mathematical structures, one very difficult to imagine given that mathematics *is* the science of structures. Also, Lakoff and Núñez do not address the indispensability argument. To do so, they would need either to reject Quine's criterion; or the thesis that mathematical physics is our best theory of the world; or the realist stance towards physical theories. Again, this is difficult as they embrace Quine's criterion themselves and seem to be realists with respect to biological theories. Finally, the intuition--based argument seems the easiest to attack from the point of view of the embodied paradigm. As we have seen above, human 'intuitive' mathematical capacities are substantially limited. However, Gödel – the proponent of the intuition-based argument – does not claim that our intuition is a faculty that gives us access to the entire world of mathematical structures. His thesis is that intuition is the source of certainty in relation to relatively simple mathematical structures and relations; more complicated mathematical statements are evaluated as true because they are justified by commonly accepted mathematical methods and have consequences controllable at the intuitive level. Of course, Lakoff and Núñez may claim that the intuition Gödel speaks of is not an intuition of *abstract* objects; it is rather the capacity to use abstract mathematical concepts, which are ultimately shaped by the experiences of our bodies. But this criticism can be softened by a modification of Gödel's argument: instead of speaking of intuition, one can simply speak of mathematical experience, even conceived of in terms of Lakoff and Núñez's theory. The crux of Gödel's argument, or so I argue, lies somewhere else: mathematical Platonism is true, because "there exist axioms so abundant in their verifiable consequences, shedding so much light upon a whole field, and yielding such powerful methods for solving problems that, no matter whether or not they are [a subject of direct experience], they would have to be accepted at least in the same sense as any well-established physical theory." Gödel points to something important here: the full power of our abstract conceptions, which lies beyond any intuitive or 'direct' experience, is clearly visible in the consequences they produce within the sphere controllable by experience, as well as in the coherence they bring to entire areas of mathematics and the heuristic role they play in solving mathematical problems. It is reasonable, therefore, to assume that those highly abstract concepts describe some independently existing structures rather than claim that they are just 'metaphorizations' of more concrete concepts. The mathematics we can somehow experience directly is only the tip of the iceberg: and when Lakoff and Núñez believe that the rest of the iceberg is only an illusion, Gödel seems to claim that it is a rock-hard, even if abstract, reality.

All this is not to say that the three arguments supporting mathematical Platonism are irrefutable or incontestable: the heated debates in the philosophy of mathematics during the last century are the evidence to the contrary. However, Lakoff and Núñez failed to provide a persuasive case against mathematical Platonism. It does not change the fact that there exists a *tension* between the embodied paradigm and mathematical Platonism: the former stresses that mathematics is *our construction*, when the latter underscores that mathematics is *independent of us*.

5. The Mathematicity of the Universe

Michael Heller introduces the concept of the mathematicity of the universe in the following words:

> In the investigation of the physical world one method has proved particularly efficient: the method of mathematical modeling coupled with experimentation (to simplify, in what follows I shall speak of the mathematical method). The advances in physics, since it has adopted the mathematical method, have been so enormous that they can hardly be compared to the progress in any other area of human cognitive activity. This incontestable fact helps to make my hypothesis more precise: the world should be ascribed a feature thanks to which it can be efficiently investigated with the use of the mathematical method. Thus the world has a rationality of a certain kind – mathematical one. It is in this sense that I shall speak of the mathematicity of the universe.[75]

According to Heller, to say that the world is mathematical is equivalent to the claim that it possesses a feature which makes the mathematical method efficient. In the quoted passage, Heller hints at one of the aspects in which the mathematicity of the world should be understood: the Efficiency Thesis. It says that the mathematicity of the universe is evident once one considers the enormous success of the mathematical method during the last 300 years. The success cannot be a pure coincidence, as the efficiency of mathematics in uncovering

---

[75] M. Heller, *Czy świat jest matematyczny?*, [in:] *idem, Filozofia i wszechświat*, Universitas, Kraków 2006, p. 48.

the laws of nature seems 'unreasonable'.[76] The argument pertaining to the 'unreasonable effectiveness of mathematics' is not trivial. As Eugene Wigner observes:

> It is true, of course, that physics chooses certain mathematical concepts for the formulation of the laws of nature, and surely only a fraction of all mathematical concepts is used in physics. It is true also that the concepts which were chosen were not selected arbitrarily from a listing of mathematical terms but were developed, in many if not most cases, independently by the physicists and recognized then as having been conceived before by the mathematicians. It is not true, however, as is so often stated, that this had to happen because mathematics uses the simplest possible concepts and these were bound to occur in any formalism. [Moreover], it is important to point out that the mathematical formulation of the physicist's often crude experience leads in an uncanny number of cases to an amazingly accurate description of a large class of phenomena. This shows that the mathematical language has more to commend it than being the only language which we can speak; it shows that it is, in a very real sense, the correct language.[77]

There are some phenomena connected to the use of the mathematical method that lead to the conclusion that it is some feature of the world that must be responsible for the method's successes. It is often the case that mathematical equations describing some aspects of the universe 'know more' than their creators. The standard story in this context is that of Einstein's cosmological constant. When Einstein formulated his cosmological equations on the basis of the newly discovered general relativity theory, he realized that they implied a dynamic, expanding universe. In order to 'stop' the expansion, he in-

---

[76] Cf. E. Winger, *The Unreasonable Effectiveness of Mathematics in the Natural Sciences*, "Communications on Pure and Applied Mathematics" 1960, vol. 13(1), pp. 1–14.
[77] *Ibid.*, p. 7.

troduced the cosmological constant. It quickly proved, however, that Einstein was 'wrong' and his equations were 'right': the expansion of the universe is a fact.

Another instructive example is given by Wigner. When Heisenberg formulated his quantum mechanics based on matrix calculus, the theory was applicable only to a few idealized problems. Applied to the first real problem, of the hydrogen atom, it also proved successful:

> This was (...) still understandable because Heisenberg's rules of calculation were abstracted from problems which included the old theory of the hydrogen atom. The miracle occurred only when matrix mechanics, or a mathematically equivalent theory, was applied to problems for which Heisenberg's calculating rules were meaningless. Heisenberg's rules presupposed that the classical equations of motion had solutions with certain periodicity properties; and the equations of motion of the two electrons of the helium atom, or of the even greater number of electrons of heavier atoms, simply do not have these properties, so that Heisenberg's rules cannot be applied to these cases. Nevertheless, the calculation of the lowest energy level of helium (...) agrees with the experimental data within the accuracy of the observations, which is one part in ten million. Surely in this case we 'got something out' of the equations that we did not put in.[78]

The second aspect of the mathematicity of the universe may be called *the Miracle Thesis*. It is possible to imagine worlds which are *mathematical* in a certain sense, yet non-idealizable. Michael Heller considers a hierarchy of such worlds. 'The most non-mathematical' is a world in which no mathematical and logical principles are observed (including any stochastic or probabilistic laws). Next, he suggests to consider a simplified model of the world: let us assume that the world in question may be in one of only two states, represented by '0' and '1'. Now:

---

[78] *Ibid.*, p. 10.

The history of this world is thus a sequence of '0's and '1's. Assume further that the world had a beginning, what may be represented by a dot at the beginning of the sequence. In this way, we get, e.g., a sequence:

.011000101011...

The task of a physicist is to construct a theory which would enable us to predict the future states of the world. Such a theory would amount to the 'encapsulation' of the sequence of '0's and '1's in a formula (which is shorter than the sequence it encapsulates). Such a formula may be found only if the sequence of '0's and '1's is algorithmically compressible. But this leads to a problem. Such a sequence may be interpreted as a decimal expansion of a number in [0,1] and – as well known – the set of algorithmically compressible numbers belonging to [0,1] is of measure 0 (…). Thus (…) there is zero-measure chance that a sequences of '0's and '1', representing our world, belongs to the set of algorithmically compressible sequences and so the physicist, who investigates such a world, may have no rational expectation to discover the theory she is looking for.[79]

This observation underscores 'the other side' of the mathematicity thesis: not only is universe mathematical (and hence penetrable by *some* mathematical method), but it is also mathematical in a non-malicious way (and hence penetrable by *our* mathematical methods).

In connection to the problem of the mathematicity of the world, Lakoff and Núñez claim:

No one observes laws of the universe as such; what are observed empirically are regularities in the universe (…); laws are mathematical statements made up by human beings to attempt to characterize those regularities experienced in the physical universe. (…) What [the physicists] do in formulating 'laws' is fit their hu-

---

[79] M. Heller, *Czy świat jest matematyczny…*, *op.cit.*, pp. 51–52.

man conceptualization of the physical regularities to their prior human conceptualization of some form of mathematics. All the 'fitting' between mathematics and physical regularities of the physical world is done within the minds of physicists who comprehend both. The mathematics is in the mind of the mathematically trained observer, not in the regularities of the physical universe.[80]

This, again, is an example of bad philosophy. Núñez and Lakoff fail to realize the far-reaching consequences of the Efficiency Thesis. What they leave unaccounted for are, at least, the fact that the mathematical method helped us to conquer the micro-scale phenomena; that equations often 'know more' than their creators; that mathematical models are often the basis for formulating qualitatively *new* predictions, and so serve as a powerful heuristic tools. It seems that behind Lakoff and Núñez's observations lies a very simplistic or naïve view of science, which rests on the *observations* of the regularities of real-world phenomena and their generalizations into the mathematically expressible laws of physics. What follows, within Lakoff and Núñez's framework one cannot even formulate the Miracle Thesis.

The interesting fact is that a conception of mathematics which draws on Lakoff and Núñez's work may shed some light on the Efficiency Thesis. The argument is quite general. Both our inborn mathematical capacities, as well as our conceptual apparatus have been shaped – in the evolutionary process – by our interaction with the environment. Now, given that our environment is mathematical (in Heller's sense of the word), it helps us to understand why our mathematical concept are efficient in uncovering the laws of the universe. Of course, such an argument cannot explain fully the efficiency of mathematics in quantum physics, or the fact that physical equations sometimes 'know more' than their creators. However, it may serve to dismiss the idea that "all the 'fitting' between mathematics and physical regularities of the physical world is done within the minds of physicists

---

[80] Lakoff, Núñez, *op.cit.*, p. 344.

who comprehend both. The mathematics is in the mind of the mathematically trained observer, not in the regularities of the physical universe." The mind is mathematical because it is a part of the mathematical universe.

6. The Triple E and the Capital M

We face a twofold problem: how to account for the necessity of mathematics on the one hand, and how to explain its efficiency in uncovering the structure of reality on the other. It seems that the 3E theory of mathematics described above offers no acceptable answers to these questions. I posit, therefore, that it should be amended, and the best way to amend it is to take advantage of Karl Popper's conception of three worlds.

> In this pluralistic philosophy the world consists of at least three ontologically distinct sub-worlds; or, as I shall say, there are three worlds: the first is the physical world or the world of physical states; the second is the mental world or the world of mental states; and the third is the world of intelligibles, or *ideas in the objective sense*; it is the world of possible objects of thought: the world of theories in themselves, and their logical relations; of arguments in themselves; and of problem situations in themselves.[81]

One can add: of mathematical relations, functions, sets, etc. I will leave aside the numerous objections raised against Popperian theory but instead I would like to concentrate on a number of its features which I believe to be essential to the discussion pertaining to the view of mathematics that is proposed here.

Firstly, world 3 really exists. One should note, however, that Popper defines existence in a special, albeit not unprecedented way: exist-

---

[81] K. Popper, *Objective Knowledge*, Oxford University Press, Oxford 1972.

ing objects have the capacity to influence one another.[82] Popper says: "The theory itself, the abstract thing itself, I regard as real because we can interact with it – we can produce a theory – and because the theory can interact with us. This is really sufficient for regarding it as real."[83] He develops this thought as follows: "[One need] only think of the impact of electrical power transmission or atomic theory on our inorganic and organic environment, or of the impact of economic theories on the decision whether to build a boat or an aeroplane",[84] in order to reject the fictitious character of the world 3 objects.

Secondly, the world 3 is autonomous. "By this – explains Popper – I mean the fact that once we have started to produce something – a house, say – we are not free to continue as we like if we do not wish to be killed by the roof falling in."[85] The autonomy of the world 3 is connected to its objectivity. Both the autonomy and the objectivity are indicated by the fact that "certain problems and relations are unintended consequences of our inventions, and that these problems and relations may therefore be said to be discovered by us, rather than invented: we do not invent prime numbers."[86] In other words, Popper indicates that any 'discovery' within the world 3 may lead to some objective consequences which are independent of our will. For example, one of the consequences of the development of Frege's logical calculus in *Begriffschrift* was the possibility of constructing the Russell paradox. Frege was unaware of this possibility; on the other hand, Russell did not invent it, he only 'discovered' it. Besides, Popper claims that the distinction between 'invention' and 'discovery' is – in most contexts – unimportant, "for every discovery is like an invention in that it contains an element of creative imagination."[87] Be that as it may, the objectivity of the world 3 is, in Popper's view, indisputable.

---

[82] *Ibid.*, p. 200.
[83] K. Popper, *Knowledge and the Body-Mind Problem*, Routledge, London 1996, p. 47.
[84] K. Popper, *Objective Knowledge…*, *op.cit.*, p. 159.
[85] K. Popper, *Knowledge and the Body-Mind Problem…*, *op.cit.*, p. 47–48.
[86] *Ibid.*, p. 48.
[87] *Ibid.*, p. 48.

Thirdly, Popper provides us with an evolutionary explanation for the emergence of the world 3. He believes – *contra* Plato – that the entities of the world 3 are not 'superhuman, divine and eternal'; they are the products of the long process of human evolution. The world 3 is a human product, in the same way as nests and dams are animal products. It is the expression of our adaptation, whose roots lie in our biology. The essential element of the evolutionary theory of the world 3 is its emergent character. Popper utilizes the classical understanding of emergence: "in the course of evolution new things and events occur, with unexpected and indeed unpredictable properties; things and events that are new, more or less in the sense in which a great work of art may be described as new."[88] Emergence leads then to new properties, which are irreducible to the properties of the underlying system.

One would be mistaken, however, to claim that the Popperian thesis that the emergent properties are irreducible is indefeasible. Although Popper stresses the implausibility of a future reductionist explanation of the emergence of life, language or mind, he confesses:

> I want to make clear that as a rationalist I wish and hope to understand the world and that I wish and hope for a reduction. At the same time, I think it quite likely that there may be no reduction possible; it is conceivable that life is an emergent property of physical bodies.[89]

This declaration is highly characteristic of Popper's ontological theory. Nowhere does he claim that the three-worlds ontology should be taken literally. He suggests only that – compared to other ontological conceptions – it is a better, more useful tool of philosophical argumentation. The world 3 is a 'useful convention'. "I would say – stresses Popper in *Knowledge and the Body-Mind Problem* – that really the name 'world 3' is just a way of putting things, and the thing

---

[88] K. Popper, J.C. Eccles, *The Self and Its Brain*, Routlege, London 1984, p. 22.
[89] K. Popper, *Objective Knowledge…*, *op.cit.*, p. 292.

is not to be taken too seriously. We can speak about it as a world, we can speak about it as just a certain region."[90] In another essay he adds:

> Whatever one may think about the status of these three worlds – I have in mind such questions as whether they 'really exist' or not, and whether world 3 may be in some sense 'reduced' to world 2, and perhaps world 2 to world 1– it seems of the utmost importance first of all to distinguish them as sharply and clearly as possible. (If our distinctions are too sharp, this may be brought out by subsequent criticism).[91]

The world 3 "is a metaphor: we could, if we wish to, distinguish more than three worlds."[92] Or: "whether or not you distinguish further regions or worlds, is really only a matter of convenience."[93] In this way, Popper tries to say that the conception of the world 3 is *a step in the right direction*. Ultimately, it may transpire that it is better to speak of one, two, or forty-seven worlds. It is crucial, however, to realize that *vis a vis* the existing ontological theories, the idea of the world 3 constitutes progress. Put otherwise: the 'division' of the reality into three worlds is *a heuristic device*. It helps us to identify *real problems* and to appreciate the role of our *theories*.

Moreover, I believe that Popper's conception should be modified in an important way. Let us observe that for Popper the emergence of the world 3 is genetically connected to the emergence of language. Without language, there would exist no such world. Thus – given the imitation-based theory of language I sketched above – one can argue that the world 3 is founded not only on what goes on in our minds, but is co-constituted by our social interactions, as it is through the social interactions that some patterns of behaviour become parts of our common cultural heritage (i.e. of the world 3). Thus, the main

---

[90] K. Popper, *Knowledge and the Body-Mind Problem…*, *op.cit.*, p. 17.
[91] K. Popper, *Unended Quest*, Routlege, London 2002, p. 211.
[92] K. Popper, *Knowledge and the Body-Mind Problem…*, *op.cit.*, p. 25.
[93] *Ibid.*, p. 18.

modification of Popper's theory I propose is the rejection of the view that the world 3 emerges from the world 2 alone. I believe that a better hypothesis reads that the world 3 supervenes both on mental states (belonging to the world 2) and social interactions (which belong to the world 1).

Let us repeat the characteristic features of the world 3: it exists in the sense of exercising influence on other objects; it is autonomous, yet we created it; and it consists of abstract objects. On this view, the tension between the fact that mathematics is independent of us, while being our creation, diminishes. In other words, I believe that the Popperian ontology provides an answer to the problem of the necessity of mathematics.

Still, on the presented view mathematics is only a part, even if a designated one, of the world 3. It means that mathematical objects are not *sui generis*; in this way, the proposed conception of mathematics escapes the argument from queerness, often raised against various kinds of Platonism. According to the argument, were there Platonic objects (e.g., values, norms, ideas) they would be queer entities, dissimilar to anything else we know. Now, when the world 3 embraces theories, values, mathematical objects, logical relations, rules, etc, they no longer are as queer as mathematical objects or values considered as *sui generis* entities.

More importantly, the claim that mathematics forms a part of the world 3 is coherent with the imitation-based view of language and culture, as so with the 3E conception of mathematics. Our culture-creating abilities, based – *inter alia* – on our imitative skills, are instrumental not only in producing mathematical knowledge, but all kinds of knowledge. Therefore, it is only justified that the products of those activities (mathematical theories, ethical directives, physical theories, etc.) fall within the same ontological category: the world 3.

The bigger problem is connected to the mathematicity of the universe. Here also, however, Popper's ontology constitutes an interesting philosophical stance. According to Popper, science, in its development, asymptotically approaches truth. Truth is, therefore, a regulatory idea of science. We will never end our scientific quest claiming that our theories are true. We would have no subjective certainty regarding their truthfulness even if they were objectively true. This means, however, that Popper is – as Stanisław Wszołek puts it – 'a transcendental essentialist':[94] he believes that the universe has a structure and our attempts at deciphering it continuously bring better results or capture some aspects of the structure of the universe. Those attempts are guided by the mathematical method, and hence the transcendental essence of the universe must be mathematical. This explains the 'unreasonable effectiveness of mathematics in the natural sciences'.

Michael Heller frames his reply in a slightly different way:

> It is obviously true that genetically our mathematics comes from the world: we abstract some of its features. However, one needs to carefully distinguish between *our mathematics* and *mathematics as such*. Our mathematics (which I also deem 'mathematics with a small m') has been developed by humans in a long evolutionary process: it is expressed in a symbolic language we invented; its results are collected in our scientific journals, books, or computer memory. But our mathematics is only a reflection of certain relations or structures, which governed the movement of atoms and stars long before biological evolution began. I deem those relations or structures mathematics as such (or 'Mathematics with a capital M'); it is what we think of when we ask, why nature is mathematical. The answer to this question, which posits that the nature is mathematical because mathematics has been abstracted from nature, turns out helpless, or even naïve, when one introduces the distinction between our mathematics and mathematics as such.[95]

---

[94] Cf. S. Wszołek, *Esencjalizm transcendentalny K.R. Poppera*, "Zagadnienia Filozoficzne w Nauce" 2002, vol. XXXI, pp. 120–132.

[95] M. Heller, *Co to znaczy, że przyroda jest matematyczna?*, [in:] *Matematyczność przyrody*, eds. M. Heller, J. Życiński, Petrus, Kraków 2010, p. 16.

Heller's claim is, then, that mathematics (with the small m) is so efficient in the process of investigating the universe, because it is a reflection of Mathematics (with the capital M), a feature of the universe. Moreover, the fact that mathematics does *resonate* with Mathematics is not as unreasonable as may seem at first: ultimately, our mathematical theories have been developed in constant interactions with the Mathematical universe.

Two things should be stressed here. Firstly, the above remarks are not intended to say that our mathematics is in a way *identical* to some part of Mathematics. The idea is only that the Popperian world 3, with our mathematical theories, is somehow capable of grasping aspects of the Mathematical universe. Secondly, even if the presented conception constitutes a philosophical reply to the Efficiency Thesis, it becomes powerless in the face of the Miracle Thesis: the universe could have been Mathematical in such a way, that the mathematics required to capture its structure would be too difficult for any carbon-based creatures to develop. This fact calls for a transcendental, if not theological reflection.

The picture resulting from the above considerations may be deemed 'The Triple E and a Capital M'. I posit the distinction between:

(1) The 'embrained' mathematics, i.e. a set of inborn basic mathematical skills (like ANS and OTS).

(2) The embodied mathematics, i.e. a conceptual system – based on our linguistic metaphorical schemes – which enables to expand human inborn mathematical capabilities.

(3) The embedded mathematics, i.e. a system of socially shared patterns of behaviour (concerning performance of various mathematical operations), propagated through imitation and enabling the stability of our mathematical knowledge.

(4) The transcendent Mathematics (with the capital M), i.e. a feature of the universe that makes possible the investigation of nature with the use of the mathematical method, as well as the evolutionary development of the mathematical brain.

This 'hierarchy' fits well the modified Popperian ontological stance: the three levels of mathematics ('embrained', embodied and embedded) give rise to our mathematical theories (with a small m), which captures some aspects of Mathematics (with the capital M). It must be stressed again that the distinction between the 'embrained', embodied and embedded mathematics (and especially between the latter two) is only analytical: we create mathematics thanks to our inborn abilities, coupled with the creative force of metaphorization enabled by language and other culture-creating skills, and sustained over time by our tendency to imitate.

Conclusion: Is the Empirical Subject Back?

To repeat: the view of mathematics defended here is that of the 3E theory: we began our mathematical journey in phylogeny, and each of us begins it in ontogeny, with some inborn mathematical skills, which are later enhanced thanks to the development of new concepts *via* metaphorization (or something akin to it) and sustained through social interactions. Our road to Cantor's paradise leads through our bodily experiences, but also through our social institutions: mathematics is, in a nontrivial sense, a joint enterprise – not only do many people contribute to the development of mathematics, but the 'small m' mathematics is co-constituted by our shared practices. Thus, one may say, there is no mathematics of the empirical subject, but rather of empirical subjects. Still, the necessity inherent in mathematics, as well as its remarkable efficiency in uncovering the structure of reality, leave us face to face with a mystery; a mystery that calls for ontological explanation and, in some cases, perhaps even for a theological one. While further advances in neuroscience and related disciplines will inevitably lead to our better understanding of what mathematical skills consist in, and – in effect – what mathematics is, some problems, such as the one summarized by the Miracle Thesis, will remain subject to insightful philosophical reflection.

Michael Heller
Copernicus Center for Interdisciplinary Studies

The Ontology of the Planck Scale[1]

Introduction

That philosophy and the sciences interact with each other, and have always done so, is a trivial fact that need not be argued for. This interaction has assumed various forms: from identification (in Antiquity and the Middle Ages), through mutual support and conflict, until open enmity. These stormy relations testify to the fact that both these human activities can hardly do without each other. In my opinion, one of the most recommended forms of their coexistence is what has been termed *philosophy in science*. It is by no means a new phenomenon. Philosophical ideas have always been present, if not in the results of scientific theories, then at least in the close context of their most fundamental notions. The phenomenon intensified with the advent of classical physics, the relativistic and quantum revolutions made it even more pronounced, and the present search for the ultimate unification of physics sometimes enhances it even beyond rational limits. Philosophical analysis is needed to filter out excessive speculations, but even what is left seems to be a "new quality" in this respect. It would perhaps be only a little exaggeration if I call it *science as philosophy*. New conceptions in theoretical physics need only the

[1] The publication of this paper was made possible through the support of a grant "The Limits of Scientific Explanation" from the John Templeton Foundation.

merest philosophical underpinning to appear as radical ontological proposals. In the present essay, I intend to illustrate this statement with a few examples taken from the current research which is trying to penetrate the physics of the Planck scale.

1. Traditional Ontologies Are Not Enough

In the history of philosophy, one can distinguish three main ontological approaches to reality;[2] let us call them, rather symbolically, the Aristotelian, the Whiteheadean and the Kantian. The latter is more epistemological than ontological, but it certainly has an ontological flavour as well.

In the Aristotelian approach, the primary entities are substances or natural bodies having certain properties that allow us to classify them into categories (bodies sharing the same property belong to the same category). Between bodies, various relations are obtained which introduce an order into the collection of bodies. Two aspects of this order are especially important: the spatial aspect and the temporal aspect. In this way, space and time from the epistemological point of view are but abstractions from the system of bodies.

In the Whiteheadean approach, reality is viewed as a process – a flux of events (not necessarily linearly ordered). Events are regarded rather as secondary with respect to the process itself, and time is not only measures the process but also belongs to its very nature.

In both these approaches, space and time can be represented in the form of a "continuum" consisting of point-like entities. This is an idealized version of the sharp localization of bodies (resp. events) in the Aristotelian (resp. Whiteheadean) approach.

The Kantian approach is different from the other two. Its primary element (although Kant himself never used this wording) is the cog-

---

[2] Since I am operating here with a very rough picture of the situation, I prefer to use the loose term "ontological approach" rather than the more technical term "ontology".

nizing subject. Space, time and other philosophical categories belong to the noetic equipment of the subject (they are its "forms *a priori*"). The world itself ("things in themselves") is not accessible to the cognizing subject. We would say that the subject projects the inner structure of its own epistemological equipment onto the world. This view, in diversified versions and various degrees of vulgarization, is quite common today. To many, it seems consonant with the prevailing fashion to emphasize the role of the human being in all aspects of life.

It goes without saying that all of these ontological approaches have been powerfully shaped by our macroscopic experience. But, as we know from contemporary physics, the macroscopic level is far from being fundamental. To construct the most adequate ontology of the world, we would have to have a "microscopic" or even "submicroscopic" experience. Of course, it is literally impossible. Our senses are unable to penetrate this level, but we have at our disposal an "intellectual experience" informed by the scientific research into the most fundamental level of the world. Based on this experience, I will face an ontological challenge coming from our insights onto what could happen at the Planck scale.

2. The Problem of the Background and Identity

All macroscopic physical theories develop in the space-time arena. Various classical theories presuppose different geometric structures for space-time, but for all of them space-time provides a "background" indispensable for identification of physical objects and for measuring or at least parametrizing physical processes. There is a common belief among theoreticians (based on some partial results obtained so far) that, from the fundamental point of view, the concept of space-time should be derived rather than given *a priori*. One of the main objections against the superstring theory is that superstrings (regarded as fundamental objects) live in a space-time background. On the other hand, proponents of the loop quantum gravity claim that their theory is "background independent" and see this as its great advantage. Perhaps the most radical in this respect are the quantum gravity and superunification theories based on noncommutative geometry and quantum groups in which, in general, the concept of point disappears.

The tendency to free physics from its rigid background finds its philosophical support in the so-called relational ontology (going back to Leibniz and even to earlier thinkers). According to this doctrine, there are no points or point-like events that are ontologically primary, but rather a network of relations, whereas points or relations are but places in this network in which it is, so to speak, especially "dense". There are, however, serious difficulties in changing these metaphorical formulations into a strict philosophical program. From the logical point of view, it is extremely difficult to get rid of objects and replace them with relations. This is well illustrated by an attempt to construct the objectless category theory. In the axioms of such a theory, the concept of object does not appear. The primary concept is that of morphism (which can be regarded as a generalized counterpart of relation) and the axioms guarantee the composition of morphisms together with its associativity and the existence of identity morphisms. It turns out, however, that objects are implicitly defended by the axioms.[3]

To overcome these difficulties, physicists invent audacious doctrines. Perhaps the most audacious one was the doctrine invented by John Archibald Wheeler who speculated that "pregeometry is the calculus of propositions";[4] in other words, that everything – the space-time background included – can be constructed out of elementary logic. Unfortunately, Wheeler's doctrine turned out an idea with no consequence for physics, but also not audacious enough. Progress in the foundations of mathematics has demonstrated that the type of logic is strictly related to the class of math-

---

[3] This is done simply by establishing the bijection between the identity morphisms and objects; see, J.P. Marquis, *From a Geometrical Point of View. A Study of the History and Philosophy of Category Theory*, Springer 2010, p. 83.

[4] See, C.W. Misner, K.S. Thorne, J.A. Wheeler, *Gravitation*, Freeman, San Francisco 1973, p. 1208.

ematical structures one employs in one's research. Boolean algebra (standard calculus of propositions) is presupposed by classical physical theories, but there are many other non-Boolean algebras,[5] and it is highly improbable that it is the Boolean algebra that governs the fundamental level. It seems that it is the type of logic that is enforced by physics rather than physics deduced from logic.

Is it impossible to eliminate objects from ontology? As we have seen, it is the identity morphism (functor, mapping) that is responsible for the existence of objects. Therefore, our question should assume the form: is it possible to construct a system without identities? We do not know the answer, but some new possibilities have been opened to our inquiry.

3. Changing Ontology

In noncommutative geometry, the concept of point is "almost a contradiction." In the usual (commutative) geometry, space can be entirely described in terms of the algebra of smooth functions defined on it. In this description, every point is identified with all smooth functions that vanish at this point. In the algebraic language, such functions form what is called a maximal ideal of the algebra of smooth functions.

The existence of this structure, apparently so "obvious", has consequences which are paramount for geometry and physics. As a result, space has local properties and admits the existence of individual objects well separated from each other. What happens here can be independent of what happens there. Stars, chairs and humans can be individual objects.

In noncommutative geometry, in general, there are no maximal ideals, and in such cases the concept of point loses its meaning.

---

[5] See, for example, D. Lambert, *Does Paraconsistent Logic Play Some Role in Physics?*, [in:] *Philosopy in Science*, eds. B. Brożek, J. Mączka, W.P. Grygiel, Copernicus Center Press, Kraków 2011, pp. 33–44.

Consequently, noncommutative spaces are in principle, nonlocal. The concept which, in a sense, replaces that of point, is the concept of state.[6] But the concept of state, as it is used in noncommutative geometry, is nonlocal. In our everyday usage, the word "state" usually also denotes nonlocal situations; for instance a crowd which can be in a state of turmoil, or in a state of idleness. In such circumstances, we must be ready to accept an ontology in which there are either no individuals (well separated from each other), or they have a nonlocal character, but what does this mean: a "nonlocal individual"?

We meet such situations in quantum mechanics in the so-called entanglement phenomena. In the EPS experiment, two electrons know about each other's spin, in spite of the fact that even the light signal cannot cover the distance separating them. Both these electrons are modelled by the same vector of the corresponding Hilbert space. This can be interpreted by saying that such a vector (an element of the Hilbert space) is a nonlocal entity, or that the two electrons have the same individuality (which is not connected to any locality). In both cases, the presupposed ontology differs from our common-sense ontology. It is worth noticing that the EPS phenomenon is a consequence of the noncommutativity of quantum mechanics. In various attempts to construct a final theory based on noncommutative geometry such effects are even enhanced.

The ontology of our microscopic world is our reference ontology. However, it is not a basic or primary ontology, but an ontology derived from an ontology of fundamental (or a "more fundamental" level). The macroscopic world is but an effect of some averaging processes constituting deeper levels of reality. This means that the ontology changes with the scale of the considered phenomena.[7] It is not given once and for all, but is a flexible concept.

---

[6] Technically, state is a linear functional on an algebra which is positive and normed to unity. There are also other concepts in noncommutative geometry that are sometimes used as analogues of points, e.g., irreducible representations of a given algebra or its characters. They are also nonlocal.

[7] I owe this idea to Dominique Lambert.

4. Change and Causality

One of the most fundamental problems of ontology is the problem of change (motion). One could risk the statement that Western ontology took off when Aristotle defended the possibility of change against Zeno's arguments. Analogously, we should now ask: how is change possible in a nonlocal environment? Of course, the concept itself of change should be suitably generalized. Noncommutative geometry provides us with such a generalization. Any motion "from point to point" is evidently excluded, but some nonlocal changes can be defined in terms of suitable mathematical structures. A good analogy is provided by the usual commutative setting in which a local change is given by a vector (e.g. a velocity vector) attached to a given point, but one can also look at the problem of motion from the global point of view by considering a vector field (that could be defined on the entire space). In noncommutative geometry there are counterparts of vector fields with the proviso that they do not consist of localized vectors.

Moreover, it is not only kinematics that can be defined in the noncommutative regime, dynamical equations can also be provided. Details depend on a particular model, but usually the dynamics is given by an operator algebra M (usually, a von Neumann algebra), and a state φ (which is a global concept) on this algebra plays here an important role. The dynamics depends on the state; if the state changes, the dynamics goes to another regime.

The problem of dynamics is strictly connected with the problem of causality. A common view (historically, attributed to Hume) that physics does not grasp the casual nexus, but only a sequence of events, is simply false. Dynamical equations not only give a sequence of events, but also model the dynamical connection between them, and it is quite natural to identify this dynamical connection with causality. In macroscopic physics, causality, understood in this way, has a local character: the cause-here and the effect-there. In the noncommutative setting, we are confronted with a generalized causality that is deprived of this local character. Taking into account the fact that the problem of causality has far-reaching ramifications for almost all branches of philosophy, we can see that the new understanding of causality could revolutionize the entire philosophical perspective.

5. Time and Probability

Time does not appear in mathematical theories; some parameters or functions can be interpreted as time only when a mathematical theory is regarded as a model of the physical world. It is rather obvious that in noncommutative geometry, when it is applied to physics, time cannot appear in the usual form as a sequence of point-like instants. Another characteristics of time is its directedness or the 'arrow of time'. In macroscopic physical theories, it is usually related to the probabilistic (statistical) properties of a given system (entropic arrow of time). If we want to look for these properties in the case of noncommutative spaces (when they are applied to physics), we should turn to their probabilistic aspects.

A generalized theory of probability, especially adapted to the noncommutative setting, does exist.[8] How can probability be defined in an environment in which there are no individuals (point-like entities)? The solution is typical for a "noncommutative thinking"; we should look for a nonlocal object which would have essential properties characteristic for the standard probabilistic measure. Such an object is a state on a given (noncommutative) algebra, and its properties essential for the probabilistic measure are its positiveness and norm equal to unity.[9] Since for a given algebra there can be many states (having these properties) defined on it, there can be many (nonequivalent) probability measures. In this sense, noncommutative probability is richer than its commutative special case (in which there is, in prin-

---

[8] See I. Cuculescu, A.G. Oprea, *Noncommutative Probability*, Kluwer, Dodrecht 1994.
[9] This generalizes the standard properties: the probability must be positive and the 'sum of all probabilities' must be equal to one. In the case of noncommutative probability, one usually also postulates that the state should be normal and faithful.

ciple, only one, Lebesgue, probabilistic measure). Technically, the 'probabilistic object' in the noncommutative probability theory is the pair (M, φ), where M is a noncommutative algebra (von Neumann algebra), and φ is a state on M. As we can see, in noncommutative geometry the probabilistic object coincides with the dynamical object. This means that every dynamics is probabilistic (in generalized sense) and every probability has a dynamic aspect.

What ontological consequences hide behind these formal results? As we can see, three important ontological properties of the macro--world are unified in noncommutative structures: time, dynamics and probability. This means that these three properties of our macroscopic world are obtained, as "limiting cases" from the "unified ontology" of the noncommutative regime. This regime has the unified temporal, dynamic and probabilistic aspects, strongly generalized as compared with the macroscopic counterparts of these aspects.[10] The generalization essentially consists of getting rid of all local properties.

What remains of probability if there is no localization (neither in space, nor in time)? The only possibility that comes to mind is a generalized propensity. In noncommutative structures, there are some tendencies or propensities which can, in principle, be computed by using noncommutative techniques, and which, in limiting processes to low energy levels, produce what we observe and measure in the macroscopic world.

The ontology of our macroscopic world is but a shadow of the ontology of the Planck scale.

---

[10] For an attempt to articulate them see: A. Connes, C. Rovelli, *Von Neumann Algebra Automorphisms and Time Thermodynamics Relation in Generally Covariant Quantum Theories*, "Classical and Quantum Gravity" 1994, vol. 11, pp. 2899–2917.

Conclusion

The long lasting success of the empirical sciences is more than a sufficient reason to treat science seriously. Since the macroscopic world is not a self-subsistent being, but an entity "derived" from more fundamental levels of the world structure, ontologies based on macro-experience (to which our every-day experiences belong) can be but approximate descriptions of what does exist. In this perspective, we should agree that ontology changes together with the progress in our penetration into the strata of the world. Since according to our current state of knowledge the most fundamental stratum is the Planck level, we should look for the deepest ontological regularities at exactly this level. Having no final reconstruction of physics at the Planck scale at present, we can only, based on partial results, speculate as to their ontological presuppositions or consequences.

In the present essay, I have attempted to present some such presuppositions or consequences related to the Planck level. They are emphasized by the majority of current models. Two properties of such "fundamental ontology" are especially striking (as compared with the "macro-ontology"), namely the generalization of concepts and their unification. Such ontologically important concepts as: dynamics (change), causality, time, space and probability, are generalized to such a degree that sometimes it is difficult to recognize in them their macroscopic counterparts. However, the latter are special cases of the former; in the sense, that in the "commutative limit" one obtains our standard notions.

The generalization is remarkable because it also leads to unification. In the process of generalization, properties that divide different concepts display more and more similar aspects.

In this view, many traditional philosophical issues should be reconsidered. To name only a few:

– The problem of God as a first cause of the universe (what sort of causality applies to the relation between God and the world?).

- The problem of the temporal beginning of the universe (what does it mean the "beginning" if there is no time on the fundamental level?).
- The problem of the human being (a new view on individuality).
- The problem of determinism and chance in the structure of the world (from the global perspective).

By asking these questions (and many others that should be asked), one can hardly doubt whether there is any philosophy in science.

Wojciech P. Grygiel
The Pontifical University of John Paul II
Copernicus Center for Interdisciplinary Studies

Spacetime in the Perspective of the Theory of Quantum Gravity
Should It Stay or Should It Go?[1]

Introduction

Space and time are common sense notions that are most immediate to how the human mind perceives and organizes the objective reality. This immediacy is so paramount that the famous German philosopher, Immanuel Kant, elevated these notions to the status of *a priori intuitions* responsible for any further sensorial experience.[2] On one hand, Kant reinforced the active role of the intellect in the process of cognition, on the other, however, he fixed the conceptual capacity to the three dimensional Euclidean framework applicable in the Newtonian mechanics thereby putting a great constraint on the possibility of the development of non-intuitive abstract mathematical concepts. This constraint was eventually overcome in the course of the formulation of the non-Euclidean geometries in the works of Nikolai Lobachevsky and Bernhard Riemann supported with Henri Poincare's idea of the *conventional* character of physical theories.[3] In commenting on

---

[1] The publication of this paper was made possible through the support of a grant "The Limits of Scientific Explanation" from the John Templeton Foundation.
[2] I. Kant, *Critique of Pure Reason*, Electronic Classics Series: Pennsylvania State University 2010, pp. 42–63.
[3] H. Poincare, *Science and Hypothesis*, The Walter Scott Publishing, London and Newcastle-on-Tyne 1905.

the influence of Mach's thought on the development of the conceptual foundations of physics, Einstein states the following:

> Concepts that have proven useful in ordering things easily achieve such an authority over us that we forget their earthly origins and accept them as unalterable givens. Thus they come to be stamped as 'necessities of thought,' 'a priori givens,' etc. The path of scientific advance is often made impassable for a long time through such errors. For that reason, it is by no means an idle game if we become practiced in analyzing the long common place concepts and exhibiting those circumstances upon which their justification and usefulness depend, how they have grown up, individually, out of the givens of experience. By this means, their all-too-great authority will be broken. They will be removed if they cannot be properly legitimated, corrected if their correlation with given things is far too superfluous, or replaced if a new system can be established that we prefer for whatever reason.[4]

The above quote well encapsulates Einstein's philosophy of physics where a physical theory is considered as a "free interplay of ideas" allowing for the evolution of theoretical concepts solely dictated by their heuristic value in explaining the observed phenomena. Concepts that are no longer useful are refined and finally replaced by ones that lead to the explanation of a larger class of phenomena. Einstein's methodology can be then rightly called *conceptual refinement*.[5] For instance, the so far separately treated notions of space and time must give in to the generalized notion of spacetime as implied by the theory of relativity. This theory is the crowning achievement of classical physics where spacetime is modelled by *the four dimensional smooth manifold with the Lorentz metric*.[6]

---

[4] A. Einstein, *Ernst Mach*, "Physikalische Zeitschrift" 1916, vol. 17, pp. 101–104.
[5] A. Fine, *The Shaky Game: Einstein's Realism and the Quantum Theory*, The University of Chicago Press, Chicago–London 1996, pp. 12–25.
[6] Cf. M. Heller, *Some Mathematical Physics for Philosophers*, Pontifical Council for Culture, Pontifical Gregorian University, Vatican, Rome 2005, pp. 27–50.

The spacetime structure in the general theory of relativity is represented by the smooth four-dimensional manifold M with the Lorentz metric γ denoted as (M, γ). Before its alterations decreed by the quantum gravity programs are discussed, a brief recap of the manifold's structure seems to be in order.[7] The most basic level is given by a set of spacetime points where the cardinal number is their only meaningful property. Devoid of any internal structure, such a set must have a topology assigned to it whereby the elements of a set have their neighbourhoods properly defined. As a result, the elements can be properly labelled and related to other elements. The next requirement is that of *differentiability (smoothness)* so that the labelling of elements can be done smoothly by assigning real numbers to the four dimensions of spacetime that characterize each of its points. This step assures that physical quantities such as position, momentum and energy that define the dynamics of a system under study can assume values out of the continuum of real numbers. As one moves one level up in the manifold, there arises another important structural layer of the manifold, namely, the *causal structure*. The imposition of causal structure assures that the transfer of information by means of electromagnetic radiation as well as slower physical signals between different regions of spacetime can be accomplished. This structure presupposes the temporal *orientability* of spacetime so that events are ordered with a cause being prior to an effect. Finally, the manifold M is equipped with the metrical structure so that distances between different events can acquire proper physical meaning.[8]

The formulation of quantum theory seriously challenged the adequacy of the spatio-temporal description. The challenge manifests itself in the famous *complementarity principle* advanced by Niels Bohr

---

[7] Cf. J. Butterfield, C. Isham, *Spacetime and the Philosophical Challenge of Quantum Gravity*, [in:] C. Callender, N. Huggett, *Physics Meets Philosophy at the Planck Scale: Contemporary Theories in Quantum Gravity*, Cambridge University Press, Cambridge 2001, pp. 77–86.

[8] For the full-fledged analysis of the global structure of spacetime see: S.W. Hawking, G.F.R. Ellis, *The Large Scale Structure of Spacetime*, Cambridge University Press, Cambridge 1973.

in the effort to retain the validity of this description within the quantum formalism.[9] Further tensions between quantum theory and the spatio-temporal description are revealed in the unification of the theory with the special theory of relativity resulting in the quantum field theory (QFT). Quantum field theory is mathematically inconsistent for it requires the application of *renormalization* procedures to eliminate divergences of the path integrals and for its accuracy it relies on the eighteen external parameters of the particle mass and charge.[10] Moreover, it must be also remembered that the smooth manifold spacetime structure as applied to the general theory of relativity is subject to the *diffeomorphism invariance* whereby the Einstein field equation is entirely independent of the selection of coordinates. The diffeomorphism invariance, however, leads to the so called *Einstein's hole problem* that casts doubt on the objective physical character of the spacetime points. Instead, the equivalence classes of certain mathematical structures are treated as ontologically fundamental.[11] Moreover, the existence of the spacetime singularities as demonstrated by Stephen Hawking and Roger Penrose suggests that the theory of general relativity is incomplete and a new theory must be developed to account for the structure of these singularities. It turns out then that the spatio-temporal description may be losing its conceptual adequacy even within the classical theories. Combined with the quantum challenge, it is not at all surprising that the majority of the researchers in the field of the theory of quantum gravity foresee the radical departure from the spacetime manifold at the most fundamental (Planck) level.

By reviewing the treatment of the manifold space-time structure in several programs of the formulating of the theory of quantum gravity, the main goal of this article is to show that the search for the quan-

---

[9] Cf. D. Murdoch, *Niels Bohr's Philosophy of Physics*, Cambridge University Press, Cambridge 1987, pp. 58 sq.

[10] Cf. L. Álvarez-Gaumé, M.Á. Vázquez-Mozo, *An Invitation to Quantum Field Theory*, Springer-Verlag, Berlin Heidelberg 2012, pp. 145 – 173.

[11] Cf. R. Rickles, F. French, *Quantum Gravity Meets Structuralism*, [in:] D. Rickles, S. French, J. Staasi, *The Structural Foundations of Quantum Gravity*, Clarendon Press: Oxford 2006, pp. 1–39.

tum structure of space-time is a complex and multifaceted task.[12] Indeed, it remains beyond doubt that some modification of the manifold structure will be necessary but due to the structure's multilayered architecture, the different programs "invade" this structure at different levels. In cases of the proposals of the theory of quantum gravity where space-time is indeed done away with, it can be recovered as an emergent entity in the appropriate low energy limits. In order to achieve the desired perspective, five most representative programs of constructing the theory of quantum gravity are surveyed. The loop quantum gravity and the superstring theory are in some sense the most conservative since they are heavily rooted in the pre-existing theories: general relativity and quantum field theory, respectively. Next, two mathematically highly sophisticated programs appear: *the twistor theory* and the *non-commutative geometries*. They can be termed more radical in the sense that they stipulate the emergence of the classical space-time from more primitive geometric and algebraic structures. Finally, however, by taking into account the results of the *topos* theory, it seems that the search for the theory of gravity may eventually go far beyond the mere continuation of the development of physics where each step of generalization demands the use of mathematical concepts of increasingly abstract and complex character and may call for the revision of the foundations of mathematics itself.

1. Loop Quantum Gravity

It is not at all surprising that researchers in the area of the theory of quantum gravity are motivated to preserve the strict rules of the treatment of space-time as implied by both constituent theories. Such

---

[12] For the detailed review of the different program of quantum gravity see for example: C. Rovelli, *Strings, Loops and Others: A Critical Survey of the Present Approaches to Quantum Gravity*, [in:] *Gravity and Relativity: At the Turn of the Millennium*, eds. N. Dadhich, J. Narlikar, Inter-University Center for Astronomy and Astrophysics, Pune, India 1998, pp. 281–331.

a program is being developed under the title of *loop quantum gravity*. Its principal founders and proponents are Abhay Ashtekar, Carlo Rovelli and Lee Smolin.[13] The general idea, however, is to propose a *quantized picture of space-time* based on the obvious fact that since space-time is a dynamical entity in general relativity, it must combine with the quantum properties of such entities in quantum mechanics resulting in space-time itself being quantized. Consequently, in the loop quantum gravity one applies the requirement that certain physical quantities such as lengths and time intervals are on one hand represented by quantum mechanical operators but on the other they must comply with the fundamental requirement of the general relativity, namely, that of the *diffeomorphism invariance* whereby the values of observables are made independent of the choice of coordinates.

In order to achieve space-time quantization, loop quantum gravity relies in its initial step on the Hamiltonian formulation of general relativity in which the four dimensional smooth manifold is decomposed into a separate 1-dimensional time coordinate that enumerates the subsequent 3-dimensional surfaces Σ of purely spatial coordinates.[14] In other words, by selecting a fixed time coordinate one obtains a *space--time foliation* where the 3-spaces Σ are slices of the foliation taken at a desired point in time. As a result, one can perform the canonical quantization by replacing physical variables with their corresponding quantum mechanical operators. This leads directly to the formulation of the famous Wheeler-DeWitt equation which is mathematically ill-defined, despite its apparent simplicity, and thus preventing its direct solution.

An important step forward was made by one of the founders of the loop quantum gravity, Abhay Ashtekar, who proposed a set of canonical variables leading to a significant simplification of the general

---

[13] For a comprehensive review of the loop quantum gravity program see for example: C. Rovelli, *Quantum Gravity*, Cambridge University Press, Cambridge 2004, p. 223 et seq.

[14] Cf. P. Mineault, *Hamiltonian Formulation of General Relativity*, McGill University 2007.

relativity theory in its Hamiltonian formulation.[15] In particular, this concerns the so called constraints imposed on the spacelike 3-surfaces Σ and their highly non-polynomial structure that acquires a tractable polynomial expression that makes it reminiscent of the quantum chromodynamics. Following the procedures of the canonical quantization, one of these constraints became the famous Wheeler-DeWitt equation. The Ashtekar canonical variables of the spacetime M for the spacelike 3-surface Σ are the components of the 3-metric γ intrinsic to Σ (momentum parameters) and the components of the spin connection Γ taken on Σ (position parameters). The results obtained by Ashtekar were further developed by Rovelli, Smolin and Jacobson, who by using the idea of the loop variables of Kenneth Wilson in the gauge theories, came up with the expressions for the loop variables in the theory of relativity.[16]

In order to obtain a fully quantum (discrete) picture of space-time, an idea of spin networks put forward by Roger Penrose was applied.[17] In regards to the origins and purpose of spin networks, Penrose writes the following:

> My own particular goal had been to try to describe physics in terms
> of discrete combinatorial quantities, since I had, at that time, been
> rather strongly of the view that physics and spacetime structure
> should be based, at root, on discreteness, rather than continuity.
> A companion motivation was a form of Mach's principle whereby
> the notion of space itself would be a derived one, and not initially
> present in the scheme. Everything was to be expressed in terms of

---

[15] A. Ashtekar, *New Variables for Classical and Quantum Gravity*, "Physical Review Letters" 1986, vol. 57, pp. 2244–7.

[16] C. Rovelli, L. Smolin, *Loop Representation for Quantum General Relativity*, "Nuclear Physics B" 1990, vol. 331, pp. 80–152.

[17] R. Penrose, *Applications of Negative Dimensional Tensors*, [in:] *Combinatorial Mathematics and Its Applications*, ed. D.J.A. Welsh, Academic Press, London 1971, pp. 221–244. For a contemporary approach see C. Rovelli, L. Smolin, *Spin Networks and Quantum Gravity*, "Physical Review D" 1995, vol. 53, 5743; arxiv: gr-qc/9505006.

the relation between objects, and not between an object and some background space.[18]

In its simplest visualization, the spin network constitutes a graph of vertices connected together by edges that correspond to the multiplies of the half-spin. In their original application by Penrose, the spin networks served as a tool to compute quantum probabilities. In the loop quantum gravity framework, however, as Rovelli points out, each vertex corresponds to a "quantum chunk of space."[19] The edges of the graph in turn represent the transverse surfaces that delimit the quanta of space. The quantized areas of these surfaces assigned to each of the edges are given the multiplies of the half-spins. In order that these spatio-temporal quantities can be in quantum mechanical terms named observables, quantum-mechnical operators must be defined for them. As is evident from the works of Rovelli and Smolin, these operators have a discrete spectra indicating that in loop quantum gravity the values of such observables as space and volume are quantized. For example, the eigenvalues of the operator of volume are given by the following equation:

$$A = 8\pi\hbar G \sum_i \sqrt{j_i(j_i + 1)}$$

where the array of numbers j, i denotes a finite sequence of half-integers.

Following the comments of Rovelli and as the discussion of the loop quantum gravity program comes to a close, it seems to make a few remarks on how the classical spacetime continuum emerges from the quantized spacetime on the Planck scale constructed with the use of spin networks. If one dimensional excitations of space are termed *weaves*, then the combination of a large number of these ex-

---

[18] R. Penrose, *The Road to Reality*, Alfred A. Knopf, New York 2005, p. 947.
[19] C. Rovelli, *Quantum Spacetime: What Do We Know?*, [in:] *Physics Meets Philosophy at the Planck Scale: Contemporary Theories In Quantum Gravity*, eds. C. Callender, N. Huggett, Cambridge University Press, Cambridge 2001, p. 110.

citations leads to the emergence of the classical spacetime.[20] Rovelli summarizes this effect in a brief but a telling comment:

> Continuous space is formed by the weave in the same manner in which the continuous two-dimensional surface of a T-shirt is formed by woven threads.[21]

2. Superstring Theory

While the loop quantum gravity program originated in the context of the theory of general relativity and, as such, does not primarily aim at the unification of all fundamental interactions, the superstring theory is a proposal that is ultimately tied with research in the area of the standard model of elementary particles. The superstring theory is currently attracting the greatest attention as a possible candidate for the fundamental theory of physics unifying all four interactions, including the consistent quantum theory of gravity. The origins of the superstring theory reach back to the works in the area of the quantum field theory of strong interactions between quarks that are known as quantum chromodynamics. It must be remembered that quantum field theory as applied to the description of particle interactions is plagued with infinities due to the singularities that occur in the computation of the corresponding Feynman path integrals and the scattering matrix S that determines the probabilities of particle trajectories. As Edward Witten, one of the major proponents of the superstring theory points out, the main advantage in replacing the point particles with one-dimensional strings is that the strings "smooth out" the singularities of the Feynman integrals thereby resolving one of the major problems

---

[20] A. Ashtekar, C. Rovelli, L. Smolin, *Weaving a Classical Geometry With Quantum Threads,* "Physical Review Letters", 1992, 69, pp. 237–240.

[21] C. Rovelli, *Quantum Spacetime: What Do We Know?,* [in:] *Physics Meets Philosophy at the Planck Scale: Contemporary Theories In Quantum Gravity,* eds. C. Callender, N. Huggett, Cambridge University Press, Cambridge 2001, p. 111.

of the standard quantum field theory.[22] Moreover, the excitations of the strings can be easily associated with elementary particles, thereby yielding a natural tool to re-produce the standard model.

The superstring theory permits the unification of all interactions including gravity but only perturbatively. This means that the theory is background-dependent (at least in its current formulation) and assumes that the strings move in a fixed space-time. Although it allows for the calculation of the scattering of a particle responsible for the gravitational interaction, the *graviton*, it is unable to explain how the dynamics space-time structure emerges out of the propagating strings. This is an obvious disagreement with the space-time ontology stipulated by the general theory of relativity.

The next objection frequently raised against superstring theory is that it is not driven by true physical motivations but solely by the criteria of mathematical consistency. Roger Penrose notes the following:

> What makes string theory so difficult to assess dispassionately is that it gains its support and chooses its directions of development almost entirely from aesthetic judgments guided by mathematical desiderata. I believe that it is important to record each of the turnings that the theory has undergone, and to point out that almost every turn has taken us further from observationally established facts. Although string theory had its beginnings in experimentally observed features of hadronic physics, it then departed drastically from those beginnings, and subsequently has had rather little guidance from observational data concerning the physical world.[23]

On one hand it is true that string theory has led to the development of certain areas of mathematics. On the other, however, the history of the theory indicates that its main advances stem from the efforts to elimi-

---

[22] E. Witten, *Reflections on the Fate of Spacetime*, [in:] *Physics Meets Philosophy...*, eds. C. Callender, N. Huggett, *op.cit.*, pp. 125–137.
[23] Cf. R. Penrose, *The Road...*, *op.cit.*, p. 888.

nate anomalies resulting in the theory's *non-renormalizability*. This mainly concerns the introduction of the extra six dimensions, supersymmetry (hence the name superstring theory) as well as the replacement of the one-dimensional strings with branes of higher dimensionality. Moreover, it is mainly due to the works of Witten that it is currently believed that the ultimate superstring theory, enigmatically named the M-theory, is a composite of several versions of the superstring theory remaining with respect to each other in the relation of duality.[24] Although the constituents theories are background-dependent, it is hoped that the final M-theory will eventually turn out to be background independent.

It is quite probable that the superstring theory may eventually offer the necessary environment for the formulation of the theory of quantum gravity and the unification of all four interactions. At this point, however, it is difficult to speculate on the eventual fate of space-time in its context. First of all, as long as the background-independent M-theory remains unknown, one cannot correctly assess the final treatment of space-time within its structure. Furthermore, the theory's incompleteness makes any ontological claims unjustified, especially in regards to the real existence of strings (or branes) at the Planck level or the objective character of the extra dimensionality beyond what is commonly accepted based on the relativity theory.

3. Non-Commutative Geometries

The *non-commutative geometries* offer an original program to formulate the theory of quantum gravity by abandoning the set-theoretical structures and moving over to richer algebraic structures. Their development is mainly due to the extensive work of a French mathematician,

---

[24] E. Witten, *String Theory Dynamics in Various Dimensions*, "Nuclear Physics B" 1995, vol. 443, pp. 85–126.

Alan Connes.[25] The algebraic approach creates a generalized mathematical environment in which individual points in spacetime acquire their meaning as notions from a more general geometric structure. In particular, this is applicable to physical theories such as the general theory of relativity in which spacetime is modelled by the four dimensional smooth manifold with the Lorentzian metric.

The entire information on the geometrical structure of the manifold, however, is encoded in the algebra A of smooth real-valued functions C^∞ (M) on the manifold M. Put simply, these functions can be mutually multiplied by themselves and by a scalar and the resulting algebra is *commutative*. Moreover, the existence of spacetime points and their environments is dependent on the existence of the maximal ideals in the C^∞ (M) algebra.[26] The manifold M can be then easily reconstructed from a given algebra A. The basic property of the quantum-mechanical operators on a Hilbert, however, is that they do not commute. It turns out that there is a possibility of constructing *a non-commutative algebra* by relaxing the requirement of the existence of the maximal ideal. Consequently, the resulting geometry whose structure is encoded by such an algebra is also non-commutative and global where the notion of a spacetime location is devoid of any meaning. This offers an excellent mathematical environment to support the quantum-mechanical notion of non-locality.

Furthermore, the non-commutative geometries seem to be a suitable mathematical structure for the unification of quantum theory with the general theory of relativity. If this path of unification turned out to be correct, it would mean the abandonment of the spatio-temporal description at the most fundamental level. Furthermore, localization in space time would be a derivative notion as one moves from the more primitive non-commutative geometry at the Planck level to the commutative regime of the classical macroworld. One can say that the non-

---

[25] A. Connes, *Noncommutative Geometry*, Academic Press, New York-London 1994.
[26] Cf. D. Perrin, *Algebraic Geometry: An Introduction*, Springer Verlag, London 2008, pp. 9–23.

-commutative geometries offer a very elegant mathematical approach to the construction of the theory of quantum gravity for they lead to mathematical structure which generalizes the geometrical approach of the general theory of relativity with the functional analysis involved in the quantum mechanical description by means of the Hilbert spaces.

The first success of the non-commutative geometries was the geometrization of the standard model of elementary particles accomplished by Connes and Lott.[27] The contemporary work conducted by Michael Heller and his collaborators aims at constructing the appropriate quantum gravitational models based on the noncommutative geometries.[28] These models do reproduce both quantum mechanics and the general theory of relativity in their appropriate limits. Interestingly enough, quantum mechanical regime is recovered by performing the measurement of an observable. In such cases, quantum mechanics acquires the status of the measurement theory.[29]

4. Twistor Theory

The twistor theory developed by Roger Penrose is a radical perspective in constructing the theory of quantum gravity that relies on his strongly metaphysical conviction that the complex numbers and their geometrical representations constitute the fundamental principle of the physical reality.[30] Although other areas of Penrose's research have received much greater attention (e.g., his theory of mind and

---

[27] A. Connes, J. Lott, *Particle Models and Noncommutative Geometry*, "Nuclear Physics B" 1990, vol. 18B, Suppl., pp. 29–47.

[28] See e.g. L. Pysiak, M. Heller, Z. Odrzygóźdź, W. Sasin, *Observables in a Non--Commutative Approach to the Unification of Quanta and Gravity: A Finite Model* 2005, "General Relativity and Gravitation", vol. 37, no. 3, pp. 541–555.

[29] M. Heller, L. Pysiak, W. Sasin, *Noncommutative Unification of General Relativity and Quantum Mechanics*, "Journal of Mathematical Physics" 2005, vol. 46, pp. 122501–16; L. Pysiak, M. Heller, Z. Odrzygóźdź, W. Sasin, *Observables in a Noncommutative Approach to the Unification of Quanta and Gravity: A Finite Model*, "General Relativity and Gravitation" 2005, vol. 37(3), pp. 541–555.

[30] A thorough exposition of the twistor theory can be found in: R. Penrose, *The Road...*, *op.cit.*, pp. 958–1009.

conscious phenomena), it is the twistor theory that bears the testimony to his deepest and unique philosophical convictions operative at the very beginning of his scientific career.[31] In particular, the theory heavily reflects his adolescent fascinations with geometry and, in particular, a certain type of stereographic projection named the Klein correspondence. In its far reaching generalization, the correspondence allows for the direct link between abstract mathematical structures and their spatio-temporal representations that conform with the categories of the human common sense cognition at the classical level. On one end, the theory is driven by the strict desiderata of the mathematical elegance and universality of the complex numbers (holomorphicity in particular), on the other, however, it seeks the union of mathematics and physics by pointing out that these particular structures offer a very natural environment for both elementary particle physics and the theory of relativity.

As Roger Penrose clearly points out, the main idea of the theory is to use the link between quantum mechanics and the spacetime structure that comes together in the Riemann sphere – the simplest complex 1-dimensional manifold that is a stereographic representation of the plane of complex numbers.[32] The usefulness of representing physical processes in its ability to represent the splitting of the field quantities into positive and negative frequencies in quantum field theory. Indeed, this is one of Penrose's initial motivations to take up the study of the twistor theory. In regards to quantum mechanics and the relativity theory, however, the sphere geometrically reflects the complex coefficients figuring into the superposition of wave vectors and the Lorentz symmetry group of the Minkowski space-time, respectively. The main assumption of the twistor theory is that space-time points (events) are derivative in regards to rays and remain with each other in relation of

---

[31] R. Penrose, *On the Origins of Twistor Theory*, [in:] *Gravitation and Geometry: a Volume in Honour of I. Robinson*, eds. W. Rindler, A. Trautman, Bibliopholis Edizioni di Filosofia e Scienze, Naples 1987, pp. 341–361.

[32] S. Hawking, R. Penrose, *The Nature of Space and Time*, Princeton University Press, Princeton 1996, pp. 109–110.

the so called *twistor correspondence*. By switching from space-time to twistor space rays Z transform into points **Z** and points R transform into Riemann spheres **R**. It is now the twistor space where the important part of physics takes place and, in particular, where the unification of quantum mechanics and relativity theory will be sought.

The part of the twistor theory that is most conceptually advanced but the most interesting is its application to the description of quantum systems. It turns out that the twistor wave functions are in themselves holomorphic, but, as Penrose states, are to be considered as elements of *the holomorphic sheaf cohomology*.[33] This is a sophisticated mathematical idea and for the purpose of this paper it will suffice to mention that the cohomology elements resemble functions defined on the twistor space Q but they exhibit fundamental non-locality. Since the cohomology element maybe shrinked to smaller regions, cohomology vanishes for regions sufficiently sized down. For instance, it can be visualized with the use of the tribar – a kind of an impossible figure, where the cohomology element is a measure of the figures impossibility. The non-locality of the spaces of the twistor functions offers a natural environment for the interpretation of quantum superpositions and the EPR effects. Despite of its mathematical elegance and sophistication, the twistor theory has not generated advances neither in the general theory of relativity nor in the theory of elementary particles.

5. Topos Theory

Although some of the presented proposals so far do suggest considerable intervention into the structure of the smooth manifold, the *topos* theory advanced and developed mainly by Christopher Isham and his co-workers is intended as an entirely new framework of constructing a physical theory in general regardless of its applications to quantum

---

[33] For an in-depth discussion of the notion of the sheaf cohomology see D. Perrin, *Algebraic Geometry: An Introduction*, Springer-Verlag London Limited 2008, pp. 113–126.

gravity.[34] On the other hand, however, quantum gravity should be naturally obtainable in such a context. The major claim of the approach is that the current physical theories are constructed with a set of a-priori and unjustified assumptions such as that of the use of the continuum of the real and complex numbers in the modelling of space-time. These assumptions can be adequately relieved in a generalized setting of a physical theory which may be considered as a kind of meta-theoretical environment. This is achieved with the use of the *topos* theory.

Before precise definitions are given, however, it seems fitting to sketch out in brief the way the topos theory proceeds in generating its frameworks to host physical theories. In the topos approach, it is only a network of logical relations that is given a-priori depending on the nature of the physical system S selected. In other words, these relations are given by the structure of a unique higher order language denoted as $L(S)$ proper to such a system. A physical theory is then understood as a *representation* of the language $L(S)$ in the topos τ_s. The notion of topos is a particular case of a category with several additional conditions imposed. In fact, the category theory and, in particular the topos theory permit the abstraction from the notion of points as elements of a set and functions between these points to a framework in which arrows constitute the basic elements and any property is established as the combinations of these arrows (morphisms). According to the formal definition, a topos is a is a Cartesian closed category with finite limits and a *subobject classifier*. A subobject classifier, in turn, is a generalization of the set-theoretical notion of the characteristic function.

The main idea of Christopher Isham that originally motivated topos theory is to construct a mathematical language in which the quantum theory could be given a more *realist* interpretation in a conceptual environment based on an *intuitionistic logic*. Quantum theory is non-realist in the sense that is set by the Kochen-Specker theo-

---

[34] C.J. Isham, J.Butterfield, *Some Possible Roles for Topos Theory in Quantum Theory and Quantum Gravity*, "Foundations of Physics" 2000, vol. 30, pp. 1707–1735.

rem according to which it is only meaningful to speak about quantum systems having properties only of the dimension of the corresponding Hilbert space is not greater than two.[35] The realist theory, on the other hand, is understood as a conceptual framework in which (1) propositions constitute a Boolean algebra and (2) they are always either true or false. In short, the main goal of the topos theory is to reconstruct the mathematical basis of the quantum theory so it looks like classical physics. The notions which have notoriously plagued the standard Copenhagen interpretation: *a measurement* and *an observer* no longer play any significant role. Since it is expected that the set-theoretical based space-time description maybe inadequate at the Planck level, the topos theory deposes points of their primary importance in favour of regions. The mathematical properties of regions are such that they do not demand any reference to points and the appropriate environments are given by the so called Heyting algebras. These algebras are generalizations of the Boolean algebras for they relax the axiom of the excluded middle as proper to aforementioned intuitionistic logic. The applicability of the topos theory is that the spatio-temporal set--theory based formalism of classical physics is an emergent structure from a more basic formalism.

Conclusion

The survey of the current programs aimed at the formulation of the theory of gravity presented above is by no means exhaustive. The discussed programs were selected in order to present the perspective of the depth of the conceptual revolution that might be necessary in order to accommodate further unification of the two disparate pictures of the microscopic and macroscopic reality generated by quantum mechanics and the relativity theory respectively. For

---

[35] C. Isham, *Lectures in Quantum Theory*, Imperial College Press, London 1995, pp. 189–219.

instance, the early program known as geometrodynamics initiated by John Wheeler[36] is based on the application of the canonical quantization of the Hamiltonian (3+1) form of the general theory of relativity and in this regard is a pre-cursor of both the loop quantum gravity discussed above and the semi-classical methods developed by Stephen Hawking.[37] The semi-classical methods are the most conservative in the space-time treatment for they fully retain the space-time continuum of the spacelike 3-surfaces that figure into the appropriate Feynman integrals. It is worthwhile to mention the group of programs resorting to the construction of the space-time models strictly from the discrete sets based on the combinatorial rules. The method known as quantum dynamical triangulation belongs to this group. The choice of the programs demonstrated in this study, however, seemed sufficient to show that the conceptual content of the mathematical structures available contemporarily in physics is large enough to obtain some genuine results leading to viable proposals of what the physical reality looks like at the Planck level. Most of the programs do agree that the space-time continuum proper to classical physics is not a suitable model at this level and a new sophisticated mathematical structure is expected to be more fundamental. The formulation of the topos theory is a definite mark that all previous programs may still carry *a priori* assumptions hidden in the mathematics itself. The final verdict in regards to the complete form of the theory of quantum gravity may involve not only physicists and mathematicians but also experts in the area of the theory of knowledge grounded in the contemporary cognitive sciences. Interestingly enough, Roger Penrose is the only quantum gravity researcher who has accepted this challenge.[38] If indeed a new mathematics is required, this will demand the further refinement of the strictest standards applicable to the human thought.

---

[36] J. Wheeler, *Geometrodynamics*, Academic Press, New York 1962.
[37] Cf. J. Hartle, S. Hawking, *Wave Function of the Universe*, "Physical Review D", 1983, vol. 28, vol. 12, pp. 2960–2975.
[38] Cf. R. Penrose, *The Emperor's New Mind*, Oxford University Press, Oxford 1989.

Helge Kragh

Aarhus University

The Criteria of Science, Cosmology and the Lessons of History[1]

Introduction

Ever since the age of Galileo, at the beginning of the Scientific Revolution, science has expanded in both breadth and depth, conquering one area after another. The development of the scientific enterprise has not occurred at a uniform growth rate, of course, but it has nonetheless been remarkably successful, progressing cognitively as well as socially and institutionally. Today, some 400 years after Galileo first demonstrated the inadequacy of the Aristotelian cosmos and the advantages of the Copernican alternative, we may wonder if there are any limits at all to scientific inquiry. Will science at some future stage enable us to understand everything? Is scientific explanation limitless? These are big questions and not really the topic of this essay, but I shall nevertheless introduce it by way of some general reflections on the limits of science, divided into four points.

(i) When it comes to the question of the limits of science, it is useful to distinguish between *knowledge* and *explanation*. After all, we may have scientific knowledge about things, even understand them on a phenomenological or instrumentalistic level, and yet be unable

---

[1] The publication of this paper was made possible through the support of a grant "The Limits of Scientific Explanation" from the John Templeton Foundation.

to provide them with an explanation. Indeed, the history of science is one long series of temporary disharmonies between phenomenal and explanatory knowledge. Early radioactivity is one example of an unexplained phenomenon that nonetheless was investigated in great detail and with great success. Another example is superconductivity, which was discovered in 1911 but only explained on a microphysical basis with the BCS (Bardeen-Cooper-Schrieffer) theory dating from 1957.

(ii) The question of scientific explanation obviously depends on our chosen criteria for what constitutes an acceptable *explanation*.[2] These criteria are not provided by nature, but by the scientific community. With an appropriate change of the criteria, scientists may be able to explain phenomena that previously seemed inexplicable. This point is particularly well illustrated by the anthropic principle, which provides explanations for a variety of phenomena – from the neutron--proton mass difference to the age of the universe – that cannot be explained on the basis of standard physics and cosmology. But are anthropic explanations proper explanations at all? As well known, this is a matter of considerable debate and a main reason why the anthropic principle is controversial.[3]

(iii) Implicitly or explicitly, the question of the limits of science refers to the problem of the *domain* of science, that is, the territory of reality to which science applies. Are there phenomena or concepts that lie outside the realm of science, or can science legitimately be applied to *all* aspects of reality? According to hard-core reductionists the latter is the case. Thus, Frank Tipler is by his own admission an "uncompromising reductionist", implying that "everything, including

---

[2] The philosophical literature on scientific and other explanations is extensive. Relevant works include R. Nozick, *Philosophical Explanations*, Harvard University Press, Cambridge, Mass. 1981, P. Achinstein, *The Nature of Explanation*, Oxford University Press, Oxford 1983, and J. Cornwell (ed.), *Explanation: Styles of Explanation in Science*, Oxford University Press, Oxford 2004.

[3] See, for example, R.J. Deltete, *What Does the Anthropic Principle Explain?*, "Perspectives on Science" 1993, vol. 1, pp. 285–305.

human beings, can be completely described by physics."[4] Generally, within the tradition of positivism the tendency has been to define reality as just those phenomena or concepts that are accessible to scientific analysis.

However, it is possible that the world that can be observed in principle (and hence be subject to scientific analysis) is only part of a larger non-physical world to which we have no empirical access and which therefore transcends the domain of science as ordinarily understood. For example, this is what has been argued within a non-theistic context by Milton Munitz, a distinguished philosopher of cosmological thought. According to him, there is a dimension of existence, which he calls "Boundless Existence", that transcends the existence of the physical universe. This Boundless Existence is not in space and time, it has no structure, and it can only be characterized – if characterized at all – in negative terms. "Boundless Existence", Munitz says, "is so totally unique (…) that all similarities with anything in our ordinary experience must fall short and be inadequate."[5]

(iv) There are questions of a conceptual nature about which we do not even know whether they are meaningful or not – or, if they are meaningful, whether they belong to the domain of science. To indicate the type of these questions, a brief reference to two problems may suffice. First, there is the much discussed question of realized or actual infinities, of whether or not there can be an infinite number of objects in the universe. The problem has become an issue in the standard inflationary model of the flat universe, but it was also discussed in relation to the earlier steady state model according to which space was infinite and uniformly populated with matter. While many modern cosmologists are perfectly happy with actual infinities, others deny their scientific legitimacy and consider the question to be

---

[4] F.J. Tipler, *The Physics of Immortality: Modern Cosmology, God, and the Renaissance of the Dead*, Doubleday, New York 1994, p. 352.

[5] M.K. Munitz, *Cosmic Understanding: Philosophy and Science of the Universe*, Princeton University Press, Princeton 1986, p. 235. See also M.K. Munitz, *The Question of Reality*, Princeton University Press, Princeton 1990.

metaphysical.[6] The point is that we do not really know whether or not it makes scientific sense. It makes mathematical and philosophical sense, but will it ever be answered scientifically?

If infinity is one of those frightening concepts on the border between physics and metaphysics, so is the concept of *nothingness* or absolute void. This is another speculation with a rich and fascinating history that has recently become relevant to science, not least after the discovery of the dark energy revealed by the acceleration of the cosmic expansion. Dark energy is generally identified with the vacuum energy density as given by the cosmological constant. However, whether or not this turns out to be true, the modern quantum vacuum is entirely different from absolute nothingness.

As far as I can see, there cannot possibly be a scientific answer to what nothingness is, and yet it does not therefore follow that the concept is meaningless.[7] Such a conclusion presupposes a rather narrow positivistic perspective.

In this essay I look at a fundamental question in the philosophy of science, namely, the defining criteria of what constitutes scientific activity from a cognitive point of view. Another and largely equivalent version of this question is the demarcation problem, that is, how to distinguish between science and non- or pseudoscience. Why is astronomy recognized as a science, when astrology and gastronomy are not? However, I shall not deal with these questions in a general and abstract way, but instead illustrate some of them by means of a couple of examples from the more recent history of cosmology. I focus on two cases, the one being the controversy related to the steady state the-

---

[6] G.F.R. Ellis, U. Kirchner, W.R. Stoeger, *Multiverses and Physical Cosmology*, "Monthly Notices of the Royal Astronomical Society" 2004, vol. 347, pp. 921–936. On the disturbing infinities appearing in steady state cosmology, see R. Schlegel, *The Problem of Infinite Matter in Steady-State Cosmology*, "Philosophy of Science" 1965, vol. 32, pp. 21–31.

[7] A useful overview is presented in R. Sorensen, *Nothingness*, "Stanford Encyclopedia of Philosophy" 2003, http://plato.stanford.edu/entries/nothingness. See also B. Rundle, *Why There Is Something Rather Than Nothing*, Oxford University Press, Oxford 2004. For the history of the concepts of vacuum and nothingness, see H. Genz, *Nothingness: The Science of Empty Space*, Basic Books, New York 1999.

ory in the 1950s and the other, the still ongoing controversy over the anthropic multiverse. Although separated in time by half a century, in some respects they are surprisingly similar and suited for comparison.

One remarkable feature shared by the two cases is the role played by philosophical considerations among the scientists themselves – philosophy *in* rather than *of* science.[8] The history of cosmology, and the history of science more generally, demonstrates that on the fundamental level philosophy is not extraneous to science but part and parcel of it. I suggest that Freeman Dyson was quite wrong when he stated, in a rare mood of positivism, that, "philosophy is nothing but empty words if it is not capable of being tested by experiments."[9] As will become clear, the views of science associated with Karl Popper's critical philosophy played an important role in both controversies. For this reason, I deal particularly with these views and Popper's emphasis on testability and falsifiability as defining criteria for science also in the area of physical cosmology. In the last section I offer some reflections on the use and misuse of historical analogies in the evaluation of scientific theories, a problem that turned up in both of the cosmological controversies.

1. Testability in the Physical Sciences

Few modern philosophers of science believe that science can be defined methodologically in any simple way and, at the same time, reflect the actual historical course of science.[10] There is no generally

---

[8] On the concept of "philosophy in science" and some of the problems related to it, see M. Heller, *How Is Philosophy in Science Possible?*, [in:] *Philosophy in Science*, eds. B. Brozek, J. Maczka, W.P. Grygiel, Copernicus Center Press, Krakow 2011, pp. 13–24.

[9] F. Dyson, *Infinite in All Directions*, Perennial, New York 2004, p. 96. A balanced argument for the value of philosophy in cosmological research is given in E. McMullin, *Is Philosophy Relevant to Cosmology?*, "American Philosophical Quarterly" 1981, vol. 18, pp. 177–189.

[10] This section relies on material discussed more fully in a paper on *Testability and Epistemic Shifts in Modern Cosmology* submitted to "Studies in History and Philosophy of Modern Physics".

accepted, more or less invariant formulation that encapsulates the essence of science and its rich variation. All the same, there are undoubtedly *some* criteria of science and theory choice that almost all scientists agree upon and have accepted for at least two centuries. Thomas Kuhn suggested five such standard criteria of evaluation, which he took to be (1) accuracy; (2) consistency, internal as well as external; (3) broadness in scope; (4) simplicity; (5) fruitfulness.[11] Although Kuhn did not mention testability as a separate criterion, it was part of the first one, according to which there must be "consequences deducible from a theory [that] should be in demonstrated agreement with the results of existing experiments and observations." Kuhn did not specifically refer to predictions, except that he included them under the notion of "fruitfulness".

Most philosophers of science, including Kuhn himself, are aware that the criteria mentioned may contradict each other in concrete situations and that a relative weighing may therefore be needed. But then the system cannot fully or uniquely determine an evaluation in a concrete case. In the context of modern cosmology, Kuhn's criteria have been discussed by George Ellis, who points out that although they are all desirable they are not of equal relevance and may even lead to conflicts, that is, to opposing conclusions with regard to theory choice.[12] Still, Ellis (and most other cosmologists) finds the first of Kuhn's criteria to be the one that in particular characterizes a scientific theory and demarcates it from other theories. In short, empirical testability is more than just one criterion out of many. Nearly all scientists consider this epistemic value an indispensable criterion for a theory being scientific: a theory which is cut off from confrontation with empirical data simply does not belong to the realm of science.

As an example, consider Einstein, who in the period from about 1905 to 1925 moved from a cautious empiricist position *à la* Mach

---

[11] T. S. Kuhn, *The Essential Tension: Selected Studies in Scientific Tradition and Change*, University of Chicago Press, Chicago 1977, pp. 321–322.
[12] G.F.R. Ellis, *Issues in the Philosophy of Cosmology*, [in:] *Philosophy of Physics*, eds. J. Butterfield, J. Earman, North-Holland, Amsterdam 2007, pp. 1183–1286.

to an almost full-blown rationalism. In his Herbert Spencer lecture of 1933 he famously stated that

> we can discover by means of pure mathematical considerations the concepts and the laws (…), which furnish they key to the understanding of natural phenomena. (…) In a certain sense, therefore, I hold it true that pure thought can grasp reality, as the ancients dreamed.[13]

But in between these two expressions of his rationalist credo, there was the no less important sentence: "Experience remains, of course, the sole criterion of the physical utility of a mathematical construction." As late as 1950, commenting on his latest attempt at a generalized theory of gravitation, he readily admitted that "Experience alone can decide on truth."[14] According to Einstein, while in the creative or constructive phase of a scientific theory empirical considerations might be wholly absent, such considerations were at the very heart of the context of justification.

While testability is universally admitted as a necessary (but not, of course, sufficient) condition for a theory to be considered scientific, in practice the concept can be interpreted in ways that are so different that the consensus may tend to become rhetorical only and of little practical consequence. The following list of interpretive questions is not complete, but it gives an idea of what physicists sometimes disagree about when it comes to testing of theories:

1. Actual testability (with present instrument technologies or those of a foreseeable future) is obviously preferable. But

---

[13] A. Einstein, *Ideas and Opinions*, Three Rivers Press, New York 1982. On Einstein's philosophy of science, see, for example, J. Shelton, *The Role of Observation and Simplicity in Einstein's Epistemology*, "Studies in History and Philosophy of Science" 1988, vol. 19, pp. 103–118, and J. D. Norton, *"Nature is the Realization of the Simplest Conceivable Mathematical Ideas": Einstein and the Canon of Mathematical Simplicity*, "Studies in History and Philosophy of Modern Physics" 2000, vol. 31, pp. 135–170.

[14] A. Einstein, *On the Generalized Theory of Gravitation*, "Scientific American" 1950, vol. 182(4), p. 17.

should it be required that a theory is actually testable, or will testability in principle – perhaps in the form of a thought experiment – suffice?

2. Should a theory result in precise and directly testable predictions, or will indirect testability do? For example, if a fundamental theory T results in several successfully confirmed predictions P_1, P_2, ..., P_n, can prediction P_{n+1} be considered to have passed a test even if it is not actually tested?[15]

3. Will a real test have to be empirical, by comparing consequences of the theory with experiments or observations, or do mathematical consistency checks also count as sufficient (theoretical) tests?

4. Another kind of non-empirical testing is by way of thought experiments or arguments of the *reductio ad absurdum* type that played an important role in the controversy over the steady state theory. A cosmological model may lead to consequences that are either contradictory or unacceptably bizarre. How should such arguments enter the overall evaluation picture?

5. At what time in the development of a theory or research programme can one reasonably demand testability? Even if a theory is not presently testable, perhaps it will be so in a future version, such as there are many examples of in the history of science.

6. How should (lack of) testability be weighed in relation to (lack of) other epistemic desiderata? E.g., is an easily testable theory with a poor explanatory record always to be preferred

---

[15] It is sometimes argued that there are reasons to believe in untestable predictions if they follow from a well-established theory with empirical success. On this account the existence of other universes is "tested" by the successfully tested background theories, in this case quantum mechanics and inflation theory. See, for example, M. Tegmark, *The Mathematical Universe*, "Foundations of Physics" 2008, vol. 38, pp. 101–150. On a different note, string theorists have suggested that the theory of superstrings has passed an empirical test because it includes gravitation without being designed to do so. E. Witten, *Magic, Mystery, and Matrix*, "Notices of the AMS" 1998, vol. 45, pp. 1124–1129.

over a non-testable theory with great explanatory power? Or what if the testable theory is overly complicated, and the non-testable one is mathematically unique and a paragon of simplicity?

7. Should predictions of novel phenomena be counted as more important than pre- or postdictions of already known phenomena? This is a question on which philosophers are divided and where the historical evidence is ambiguous.

2. A Historical Case: The Steady State Theory

The steady state theory of the universe, proposed by Fred Hoyle, Hermann Bondi and Thomas Gold in 1948, aroused a great deal of philosophical interest, in part because of the theory's controversial claim of the continual creation of matter and more generally because of its appeal to philosophy and methods of science. For example, Bondi and Gold argued that the new steady state theory was preferable from a methodological point of view, as it was simpler, more direct, and more predictive than the cosmological theories based on general relativity. The latter class of theories, they said, was "utterly unsatisfactory" since it covered a whole spectrum of theories that could only be confronted with the observed universe if supplied with more or less arbitrary assumptions and parameters:

> In general relativity a very wide range of models is available and the comparisons [between theory and observation] merely attempt to find which of these models fits the facts best. The number of free parameters is so much larger than the number of observational points that a fit certainly exists and not even all the parameters can be fixed.[16]

---

[16] H. Bondi, T. Gold, *The Steady-State Theory of the Expanding Universe*, "Monthly Notices of the Royal Astronomical Society" 1948, vol. 108, p. 269 and p. 262.

Relativistic cosmology sorely lacked the deductive character of the steady state theory, which uniquely led to a number of predictions, such as the mean density of matter, the curvature of space, and the average age of galaxies. According to Bondi and Gold, the predictions were crucially based on what they called the perfect cosmological principle (PCP), namely, the postulate that there is neither a privileged place nor a privileged time in the universe. Thus, the PCP is a temporal extension of the ordinary cosmological principle (CP).

Whether in the Bondi-Gold or the Hoyle version, the steady state theory was critically discussed by many philosophers and philosophically minded astronomers and physicists.[17] To the first category belonged Adolf Grünbaum, Mario Bunge, Milton Munitz, Norwood Russell Hanson, and Rom Harré, and to the latter Herbert Dingle, Gerald Whitrow, William McCrea, and William Bonnor. We witness in this discussion an instructive case of philosophy in science, an unusual dialogue between professional philosophers and the spontaneous philosophy of practicing scientists.

Much of the methodological discussion in the 1950s and 1960s focused on the criteria on which to judge the scientific nature of the steady state theory, or of cosmology in general. To give just a couple of examples, Dingle found the cosmological principle – whether in its original CP or the "perfect" PCP form – to be plainly unscientific because it was *a priori* and hence in principle inviolable.[18] According to Bunge and some other critics, the steady state theory was

---

[17] On the philosophical foundation of steady state cosmology and the discussion of its scientific status, see Y. Balashov, *Uniformitarianism in Cosmology: Background and Philosophical Implications of the Steady-State Theory*, "Studies in History and Philosophy of Science" 1994, vol. 25, pp. 933–958, and H. Kragh, *Cosmology and Controversy: The Historical Development of Two Theories of the Universe*, Princeton University Press, Princeton 1996.

[18] H. Dingle, *Cosmology and Science*, [in:] *The Universe*, eds. G. Piel et al., Simon and Schuster, New York 1956, pp. 131–138. The misguided claim that the cosmological principle is *a priori* has more recently been made by the German philosopher Kurt Hübner, according to whom cosmological models rest on *a priori* constructions that are essentially independent of observations. K. Hübner, *Critique of Scientific Reason*, University of Chicago Press, Chicago 1985, pp. 150–152.

nothing but "science-fiction cosmology" because it rested on the ad hoc assumption of continual creation of matter.[19] On the other hand, and contrary to the later multiverse controversy, testability was not at the heart of the discussion. Both parties accepted that a cosmological theory must be observationally testable, but they rated this epistemic value somewhat differently and did not draw the same conclusions from it.

In 1954 Bondi and Whitrow engaged in an interesting public debate concerning the scientific status of physical cosmology. Whitrow, stressing the unique domain of cosmology, argued that it was not truly scientific and probably never would be so. It would remain, he thought, a borderland subject between science and philosophy. Bondi, on the other hand, suggested that the hallmark of science was falsifiability of theories and that on this criterion cosmology was indeed a science. "Every advocate of any [cosmological] theory will always be found to stress especially the supposedly excellent agreement between the forecasts of his theory and the sparse observational results", he admitted. And yet,

> The acceptance of the possibility of experimental and observational disproof of any theory is as universal and undisputed in cosmology as in any other science, and, though the possibility of logical disproof is not denied in cosmology, it is not denied in any other science either. By this test, the cardinal test of any science, modern cosmology must be regarded as a science. (…) I consider universal acceptance of the possibility of experimental disproof of any claim an absolute test of what constitutes a science.[20]

Although not mentioning Karl Popper by name, Bondi was clearly defending a main methodological point in Popperian philosophy which

---

[19] M. Bunge, Cosmology and Magic, "The Monist" 1962, vol. 47, pp. 116–141.

[20] G.J. Whitrow, H. Bondi, *Is Physical Cosmology a Science?*, "British Journal for the Philosophy of Science" 1954, vol. 4, p. 279 and p. 282. For the Bondi-Whitrow discussion, see also H. Kragh, *Cosmology and Controversy, op.cit.*, pp. 233–237.

he much admired. Whitrow, who was also well acquainted with Popper's views, did not disagree, although he warned that falsifiability should not be considered a final and absolute criterion: "The important role of disproof in science, which has been so cogently argued by K. R. Popper, is intimately related to the self-correcting tendency of science and this, in my view, is another aspect of the pursuit of unanimity."[21]

Although Popperian criteria of science played a considerable role during the cosmological controversy, and were highlighted by the steady state proponents in particular, they were rarely an issue of dispute. By and large, criteria of a Popperian kind were accepted also by many cosmologists favouring an evolving universe governed by the laws of general relativity. One of them was the British astronomer George McVittie, who was strongly opposed to the steady state theory and other theories he suspected were based on *a priori* principles. He described the philosophical foundation of the Bondi-Gold theory as "Karl Popper's dictum that a scientific theory can never be proved to be true but, instead, that certain theories can be proved to be false by an appeal to observation." While he considered the dictum to be a "probably unimpeachable doctrine", he parodied Bondi's use of it. If one followed Bondi's vulgar version of Popper's philosophy, "we should be justified in inventing a theory of gravitation which would prove that the orbit of every planet was necessarily a circle. The theory would be most vulnerable to observation and could, indeed, be immediately shot down."[22]

---

[21] G.J. Whitrow, H. Bondi, *Is Physical Cosmology a Science?*, op.cit., p. 280.

[22] G.C. McVittie, *Rationalism versus Empiricism in Cosmology*, "Science" 1961, vol. 133, p. 1231. McVittie belonged to what he called the "observational school" in cosmology. See J.-M. Sánchez-Ron, *George McVittie, the Uncompromising Empiricist*, [in:] *The Universe of General Relativity*, eds. A.J. Kox, J. Eisenstaedt, Birkhäuser, Boston 2005, pp. 189–222.

3. A Modern Case: The Anthropic Multiverse

Like the earlier controversy over the steady state cosmological model, the present discussion of the multiverse hypothesis deals to a large extent with philosophical issues and the borderline between science and philosophy.[23] Both cases are about foundational issues that cannot be answered simply by observation and calculation. Among those issues are: Does the theory belong to science proper, or is it rather a philosophical speculation? If it disagrees with established standards of science, should these standards perhaps be changed? What are the basic criteria for deciding whether a theory is true or false? The discussion in 2008 between Bernard Carr and George Ellis concerning the multiverse, taking place in the journal *Astronomy & Geophysics*, can be seen as a modern analogue of the 1954 Bondi-Whitrow discussion about the scientific nature of physical cosmology.[24]

However, although the two cosmological controversies have enough in common to make a comparison meaningful, there are also some dissimilarities. As mentioned, in the case of the steady state theory there was a great deal of interest from the side of the philosophers, who were key players in the debate. Strangely, a corresponding interest is largely absent in the case of the multiverse debate, where the philosophically related questions are predominantly discussed by the physicists themselves. Another difference is that the overarching question of the multiverse hypothesis is whether or not it is testable by ordinary observational means. Does it result in predictions of such a kind that, should they turn out to be wrong, the hypothesis must be

---

[23] The central source in the multiverse debate is B. Carr (ed.), *Universe or Multiverse*, Cambridge University Press, Cambridge 2007. See also H. Kragh, *Higher Speculations: Grand Theories and Failed Revolutions in Physics and Cosmology*, Oxford University Press, Oxford 2011, where further references are given. More popular accounts of the multiverse (in one or more of its several versions) include L. Susskind, *The Cosmic Landscape: String Theory and the Illusion of Intelligent Design*, Little, Brown and Company, New York 2006, and A. Vilenkin, *Many Worlds in One: The Search for Other Universes*, Hill and Wang, New York 2006.

[24] B. Carr, G.F.R. Ellis, *Universe or Multiverse?*, "Astronomy & Geophysics" 2008, vol. 49, pp. 2.29–2.37.

wrong as well? In this respect, the cases of the steady state and the multiverse are quite different: whereas the first was eminently falsifiable – and was in fact falsified – the multiverse fares very badly in terms of falsifiability. As has often been pointed out, it explains a lot but predicts almost nothing.

The current discussion concerning the multiverse involves two major questions of obvious relevance to the philosophy of and in science:

(i) Has cosmology become truly and exclusively scientific, in the sense that philosophical considerations no longer play a legitimate role? If so, has it achieved this remarkable status by changing the rules of science?

(ii) Which people or groups have the "right" to define these rules of science and thus to decide whether or not a particularly theory discussed by the scientists is in fact scientific?

It is far from clear whether some of the recent developments, such as multiverse cosmology and aspects of so-called physical eschatology, belong primarily to science or philosophy. The idea of many universes, traditionally a subject of philosophical speculation, is now claimed to have been appropriated by the physical sciences. Is this yet another conquest of the ever-victorious physics, at the expense of philosophy? According to Max Tegmark, this is indeed the case. "The borderline between physics and philosophy has shifted quite dramatically in the last century", he comments. "Parallel universes are now absorbed by that moving boundary. It's included within physics rather than metaphysics."[25] However, sceptics disagree.

One problem with the multiverse hypothesis is that the excessive amount of universes seems to allow almost any physical state of affairs – if not in our universe, then in some other. This, together with the unobservability of the other universes, tends to make the multiverse unacceptable from Popperian-like points of view. According to Popper's philosophy, a scientific theory must be falsifiable and there-

[25] Quoted in C. Seife, *Physics Enters the Twilight Zone*, "Science" 2004, vol. 305, p. 465.

fore set constraints to the results of possible observations: "Every 'good' scientific theory is a prohibition: it forbids certain things to happen", as he said in a lecture of 1953.[26] At least in some versions, multiverse cosmology suffers from an extreme lack of prohibitiveness.

Some physicists advocating the multiverse and anthropic reasoning have questioned whether there is any need for external norms of science of a philosophical nature, these norms being Popperian or something else. "If scientists need to change the borders of their own field of research", says the French cosmologist Aurélien Barrau, "it would be hard to justify a philosophical prescription preventing them from doing so."[27] Leonard Susskind, the leading advocate of the string-based landscape multiverse theory, agrees with Barrau that normative prescriptions are unnecessary and may even be harmful. He suggests that only the scientists themselves, or perhaps their scientific communities, can decide by means of their practices what is and what is not science: "It would be very foolish to throw away the right answer on the basis that it doesn't conform to some criteria for what is or isn't science."[28] Susskind is particularly dissatisfied with the falsification criterion and what he calls the "overzealous Popperism" advocated by the "Popperazi" following Popper's philosophy. "Throughout my long experience as a scientist", he says, "I have heard unfalsifiability hurled at so many important ideas that I am inclined to think that no idea can have great merit unless it has drawn this criticism. (…) Good scientific methodology is not an abstract set of rules dictated by philosophers."[29]

It needs to be pointed out that the Barrau-Susskind argument is deeply problematic and hardly tenable. Not only is it circular reasoning to define science as what scientists do, it also presupposes that all

---

[26] K.R. Popper, *Conjectures and Refutations*, Routledge, New York 1963, p. 48.

[27] A. Barrau, *Physics in the Universe*, "Cern Courier" 2007 (20 November, online edition).

[28] Quoted in G. Brumfiel, *Outrageous Fortune*, "Nature" 2006, vol. 358, p. 363.

[29] L. Susskind, *The Cosmic Landscape, op.cit.*, pp. 193–195. See also H. Kragh, *Higher Speculations, op.cit.*, pp. 280–285.

scientists have roughly the same ideas of what constitutes science, which is definitely not the case. Not even within such a relatively small field as theoretical cosmology is there any consensus. Subjects that scientists find interesting and discuss at conferences or write articles about in peer-reviewed journals do not automatically belong to the realm of science. Moreover, it makes no sense to speak of a "right answer" without appealing, explicitly or implicitly, to some criteria of science. To conclude that a theory is either valid or invalid necessarily involves certain standards of scientific validity. These standards need not be part of a rigid philosophical system ("dictated by philosophers"), nor do they have to be explicitly formulated, but it is hard to see how they can be avoided. Nature herself does not provide us with the criteria for when an answer is right.

4. Karl Popper and Modern Cosmology

As already indicated, Popper's philosophy of science has played, and continues to play, an important role in methodological debates concerning cosmology. According to a study by Benjamin Sovacool, astronomers and cosmologists often invoke Popper's ideas as a guide for constructing and evaluating theories, although they rarely reveal a deeper familiarity with these ideas.[30] The first time Popperianism entered the scene of cosmology was in the 1950s, in connection with the steady state theory and Bondi's explicit use of standards based on Popper's philosophy of science. In a discussion of modern cosmology from 1960, he summarized Popper's view as follows:

> The purpose of a theory is to make forecasts that can be checked
> against observation and experiment. A scientific theory is one that
> it is in principle possible to disprove by empirical means. It is this

---

[30] B. Sovacool, *Falsification and Demarcation in Astronomy and Cosmology*, "Bulletin of Science, Technology & Society" 2005, vol. 25, pp. 53–62.

supremacy of empirical disproof that distinguishes science from other human activities. (...) A scientific theory, to be useful, must be testable and vulnerable.[31]

The leading theoretical physicist and cosmologist Lee Smolin is no less a "Popperazo" than Bondi was. As Bondi used Popper's philosophy to criticize the big bang theory, so Smolin uses it to dismiss most versions of multiverse cosmology. "According to Popper," he says, "a theory is falsifiable if one can derive from it unambiguous predictions for practical experiments, such that – were contrary results seen – at least one premise of the theory would have been proven not true. (...) Confirmation of a prediction of a theory does not show that the theory is true, but falsification of a prediction can show it is false."[32]

In regard of the considerable impact of Popper's thoughts, it is remarkable that physical cosmology is hardly mentioned at all in his main works. Yet a closer look reveals that cosmology does turn up in his books and papers, most explicitly in a lecture given in 1982 in Alpbach, Austria. Calling cosmology "the most philosophically important of all the sciences," at this occasion he praised the by then defunct Bondi-Gold-Hoyle theory as "a very fine and promising theory", not because it was true but because it was testable and had in fact been falsified. As a result of measurements based on methods of radio astronomy, "it seems to have been refuted in favour of the (older) big bang theory of expansion."[33] Popper did not mention the cosmic microwave background radiation or other evidence (such as the measured amount of helium in the universe) that had laid the steady state theory in the grave.

---

[31] H. Bondi, *The Steady-State Theory of the Universe*, [in:] *Rival Theories of Cosmology*, eds. H. Bondi et al., Oxford University Press, London 1960, p. 12.

[32] L. Smolin, *Scientific Alternatives to the Anthropic Principle*, [in:] *Universe or Multiverse?*, ed. B. Carr, Cambridge University Press, Cambridge 2007, pp. 323–324. Emphasis added. For Smolin as a self-declared "Popperazo" see L. Smolin, *The Trouble with Physics*, Penguin Books, London 2008, p. 369.

[33] K.R. Popper, *In Search of a Better World: Lectures and Essays from Thirty Years* 1994, Routledge, London, pp. 58–60.

Although references to Popper's philosophy of science often appear in modern cosmology, it is probably fair to say that few physicists and astronomers have actually read him. Most seem to rely on what they have been told or happen to know from the secondary literature. This results in discussions that are sometimes simplistic and based on misunderstandings. What cosmologists (and other scientists) discuss is most often naïve falsificationism rather than the sophisticated versions of authentic Popperianism.[34] Popper's views, including his awareness that falsifiability cannot stand alone as a demarcation criterion, were far from the caricatures one can sometimes meet in the science literature. It should be recalled that his philosophy was normative and that he did not claim that the associated standards reflected the actual practice of scientists. Moreover, he never held that falsifiability is a sufficient condition for a theory being scientific, but only that it is a necessary condition. Although somewhat ambiguous with regard to the relationship between his methodological rules and scientific practice, he admitted that strict falsifiability does not belong to the real world of science:

> In point of fact, no conclusive disproof of a theory can ever be produced; for it is always possible to say that the experimental results are not reliable, or that the discrepancies which are asserted to exist between the experimental results and the theory are only apparent and that they will disappear with the advance of our understanding. (…) If you insist on strict proof (or strict disproof) in the empirical sciences, you will never benefit from experience, and never learn from it how wrong you are.[35]

---

[34] As pointed out in M. Heller, *Ultimate Explanations of the Universe*, Springer-Verlag, Berlin 2009, pp. 88–89.

[35] K.R. Popper, *The Logic of Scientific Discovery*, Basic Books, New York 1959, p. 50. In a note appended to the English edition, Popper remarked that "I have been constantly misinterpreted as upholding a criterion (and moreover one of *meaning* rather than of *demarcation*) based upon a doctrine of 'complete' or 'conclusive' falsifiability."

Contrary to what many scientists believe, Popper did not assign any absolute value to the criterion of falsifiability and did not consider it a *definition* of science. He recognized that the distinction between metaphysics and science is often blurred. "What was a metaphysical idea yesterday can become a testable theory tomorrow," he wrote.[36] Far from elevating falsificationism to an inviolable principle, he suggested that it is itself fallible and that it may be rational to keep even an admittedly wrong theory alive for some time:

> There is a legitimate place for dogmatism, though a very limited place. He who gives up his theory too easily in the face of apparent refutations will never discover the possibilities inherent in his theory. *There is room in science for debate*: for attack and therefore also for defence. Only if we try to defend them can we learn all the different possibilities inherent in our theories. As always, science is conjecture. You have to conjecture when to stop defending a favourite theory, and when to try a new one.[37]

This is indeed a view far from the strict or naïve falsificationism often discussed by scientists either for or against Popper. It is a view closer to the one associated with philosophers of science such as Imre Lakatos and Thomas Kuhn.

5. The Role of Historical Analogies

Just like scientists use methodological and other philosophical arguments in evaluating the value of a fundamental scientific theory, sometimes they use (or misuse) arguments relating to the history of science. The typical way of doing this is by supporting an argument

---

[36] K.R. Popper, *Replies to my Critics*, [in:] *The Philosophy of Karl Popper*, ed. P.A. Schilpp, Open Court Publishing House, La Salle, IL 1974, p. 981.
[37] K.R. Popper, *Replies to my Critics*, *op.cit.*, p. 984. Popper's emphasis.

of a philosophical kind by means of concrete historical cases in the form of exemplars. That is, history is used analogically. The standard formula is this: Since, in a certain historical case, the epistemic value x proved successful, a modern theory should preferably incorporate x; or, conversely, if values of the kind y have proved a blind alley in the past, they should be avoided in a modern theory. The values or prescriptions x and y will usually be those associated with either well known successes or failures in the history of science. Often it is enough to associate them with the great authorities of the past.

Historical analogy arguments of this kind are quite common in controversies and in discussions of theories of a foundational nature. Einstein often relied on historical exemplars when he wanted to illustrate and give support to his favourite deductivist methodology of science, such as he did in the semi-popular book *The Evolution of Physics*.[38] During the cosmological controversy in the 1950s, some physicists and astronomers used Galileo's supposed empiricism as a weapon against what they considered rationalistic and *a priori* tendencies in the steady state theory. McVittie associated this theory with Aristotle's dogmatic world system (!) and the empirical cosmology based on general relativity with Galileo's physics. Dingle similarly claimed that the perfect cosmological principle has "precisely the same nature as perfectly circular orbits and immutable heavens" and that "it is largely identical with the Aristotelian principle of celestial regions."[39] It was, and still is, common to refer to the epicycles of ancient astronomy when scientists want to criticize a theory for being complicated and ad hoc.

In other cases, the references to history are not to concrete events or persons, but of the "history suggests" type where the record of some general idea in past science is used to evaluate the methodo-

---

[38] For an analysis of Einstein's attitude to and use of the history of science, see H. Kragh, *Einstein on the History and Nature of Science*, [in:] *The Way through Science and Philosophy*, eds. H.B. Andersen *et al.*, College Publications, London 2006, pp. 99–118.
[39] H. Dingle, *Cosmology and Science, op.cit.*, p. 137.

logical basis of a modern theory. For example, string theory notoriously lacks connection to experiment and is, according to some critics, largely justified by the dubious idea that fundamental physics must be mathematically beautiful. One of the critics, Daniel Friedan, says: "History suggests that it is unwise to extrapolate to fundamental principles of nature from the mathematical forms used by theoretical physics in any particular epoch of its history, no matter how impressive their success. (…) Mathematical beauty in physics cannot be appreciated until after it has proved useful."[40]

Again, although the anthropic principle does not lead to precise predictions, it may be justified by referring to historical cases in which a theory has been highly successful in spite of its limited predictivity. The prime example of such a theory is Darwinian evolution, which is sometimes referred to in the debate over the standards to be used in fundamental physics and cosmology. "One is reminded of Darwin's theory, which is a powerful explanatory tool even though some question its predictive power," says Craig Hogan. "Anthropic arguments are vulnerable in the same way to 'Just So' storytelling but may nevertheless form an important part of cosmological theory."[41]

One historical case that occurs surprisingly often in the universe--or-multiverse discussion is Kepler's geometrical model of the heliocentric universe as expounded in his *Mysterium Cosmographicum* from 1596. When multiverse proponents refer to Kepler's model, it is invariably as a negative exemplar, to illustrate that the universe is probably not uniquely described by the mathematical solutions to the equations of physics. According to Steven Weinberg, "We may just have to resign ourselves to a retreat, just as Newton had to give up Kepler's hope of a calculation of the relative sizes of planetary orbits from first principles."[42] Frank Wilczek uses the same case to argue for

[40] D. Friedan, *A Tentative Theory of Large Distance Physics*, "Journal of High Energy Physics" 2003, vol. 10, p. 63.
[41] C.J. Hogan, *Why the Universe Is Just So*, "Reviews of Modern Physics" 2000, no. 72, p. 1160.
[42] S. Weinberg, *Living in the Multiverse*, [in:] *Universe or Multiverse, op.cit.*, p. 39.

the same conclusion: "In the development of Copernican-Newtonian celestial mechanics, attractive *a priori* ideas about the perfect shape of planetary orbits (Ptolemy) and their origin in pure geometry (Kepler) had to be sacrificed."[43] On the other hand, Kepler may also be used as a positive exemplar (and Galileo as a negative exemplar), as Martin Rees does in his argument for the multiverse: "Kepler discovered that planets moved in ellipses, not circles. Galileo was upset by this. (…) The parallel is obvious. (…) A bias in favour of 'simple' cosmologies may be as short-sighted as was Galileo's infatuation with circles."[44]

My last example of the questionable use of history of science comes from Carr, who suggests that critics of the multiverse are on the wrong side of history. Throughout the history of cosmology, the universe has always been conceived as bigger and bigger, he claims, so why be satisfied with a single universe instead of a whole lot of them? Carr's argument may have some rhetorical force, but it is poor from both the perspective of history and from a logical point of view. At any rate, here it is:

> Throughout the history of science, the universe has always gotten bigger. We've gone from geocentric to galactocentric. Then in the 1920s there was this huge shift when we realized that our galaxy wasn't the universe. I just see this as one more step in the progression. Every time this expansion has occurred, the more conservative scientists have said, 'This isn't science.' This is the same process repeating itself.[45]

This is not the place for discussing the role of history of science in scientific or philosophical arguments, but it needs to be pointed out

---

[43] F. Wilczek, *Enlightenment, Knowledge, Ignorance, Temptation*, [in:] *Universe or Multiverse*, *op.cit.*, p. 50.

44 M. Rees, *Explaining the Universe*, [in:] *Explanation*, *op.cit.*, p. 63.

[45] Quoted in T. Folger, *Science's Alternative to an Intelligent Creator: The Multiverse Theory*, "Discover Magazine" (online version) 2008. In fact, the universe has not "always gotten bigger." Kepler's universe was much smaller than Copernicus's, and Kant's universe of the 1750s was much bigger than the Milky Way universe a century later.

that, in general, one should be very cautious with reasoning based on historical analogies and extrapolations from historical trends. Historical arguments and analogies have a legitimate function in the evaluation of current science.[46] We cannot avoid being guided by the past, and it would be silly to disregard the historical record when thinking about the present and the future. On the other hand, such guidance should be based on historical insight and not, as is often the case, on arbitrary selections from a folk version of history. Generally speaking, the history of science is so diverse and complex that it is very difficult to draw from it lessons of operational value for modern science. In 1956, in connection with the controversy over the steady state theory, Gold reflected on the lessons of history of science with regard to the methodology of cosmology and other sciences. He considered history to be an unreliable guide:

> Analogies drawn from the history of science are frequently claimed to be a guide [to progress] in science; but, as with forecasting the next game of roulette, the existence of the best analogy to the present is no guide whatever to the future. The most valuable lesson to be learned from the history of scientific progress is how misleading and strangling such analogies have been, and how success has come to those who ignored them.[47]

Of course, scientists should not ignore history. They can and should use the rich treasure of resources hidden in the history of science, but they must do it with the proper caution and professional insight.

---

[46] L. Darden, *Viewing the History of Science as Compiled Hindsight*, "AI Magazine" 1987, vol. 8:2, pp. 33–41; H. Kragh, *An Introduction to the Historiography of Science*, Cambridge University Press, Cambridge 1987, pp. 150–158.
[47] T. Gold, *Cosmology*, "Vistas in Astronomy" 1956, vol. 2, p. 1722.

Bogdan Dembiński
University of Silesia

Structuralism in the Platonic Philosophy of Science[1]

In the contemporary disputes concerning the status and comprehension of the philosophy of science, we may encounter numerous references to the whole of philosophical tradition. One of them, the tradition of Platonic philosophy, is undoubtedly very significant. Its presence is particularly visible in the area of natural science. It is hard to imagine discussions and arguments within its scope without reference to Platonism as broadly defined. It generates a natural curiosity and interest in the conception of Platonic philosophy of science. In the time context in which Plato wrote, one should rather talk about Platonic philosophy of knowledge. This problem is called as such in the dialogue *Theaetetus*.[2] Plato asks in it: what is knowledge and in what ways can it be achieved? He assumes it should be defined as *alethes doksa metha logou*, which might be translated as: true belief confirmed by reason. The content of the analyses presented is an attempt to determine the sense of this definition. At the same time, it is an attempt to reach the essence of Platonic conception of science and its realisation which we deal with in philosophy of mathematics.

---

[1] The publication of this paper was made possible through the support of a grant "The Limits of Scientific Explanation" from the John Templeton Foundation.
[2] Plato, *Theaetetus*, 146a, [in:] *Plato in Twelve Volumes*, translated by H.N. Fowler, Harvard University Press, Cambridge, MA; William Heinemann Ltd., London 1925. Hereafter, all citations refer to this translation of Plato's works.

I propose to begin this analysis from the dialogue in which the most general conditions of philosophical discourse are established and Plato assumes them in his philosophy, foremost in the *Parmenides* dialogue. Plato claims that the fundamental issue which conditions philosophical discourse is the issue of the comprehension of the relation between one and multiplicity. The thing is that any form of being becomes the subject of consideration and its analysis ultimately focuses on the question of in what way is it one and in what way multiplicity? In the hierarchy of questions, one cannot ascend any higher. Is one merely the one created from some multiplicity or one exists and may exist independently, irrespective of multiplicity? While applying the contemporary language of logic one should ask: are we dealing with the collective or distributive comprehension of one? In the first case, one would be solely the sum of the parts which make up this one, whereas, in the second case, one would be an independent wholeness transcending beyond its constituent parts. Plato is in favour of the second solution, arguing that the whole is something more than the sum of its constituent parts. It should be added that such a whole is the whole owing to which parts only then obtain their full determinacy. The reason for this state of affairs is viewed by Plato in the act of participation of one in being (*hen ousia metehein*). One, provided it participates in being, generates the whole which is not a simple sum of parts. In the first hypothesis in the dialogue[3] Plato considers what would one be, comprehended as whole, which would not participate in being, thus it would be a completely isolated one. He remarks that the autonomous and isolated one would be unknowable and it could not make up any whole entitled to any parts. While we are to analyse beings available to our cognition (it may be a real, ideal or mental being) they must consist of some one and multiplicity. In Plato's language it means that one, while participating in being (thereby it ceases to be the same isolated one), develops into multiplicity. One, Plato says, if it is (participates in being) then it is and it

---

[3] Plato, *Parmenides*, 137c–142a.

is one. One has in itself that it is and is has in itself that it is one. One while participating in being develops into infinite multiplicity (*apeiron plethos*).[4] This multiplicity is a necessary result of participation (*metheksis*) of one in being. In this act, one also "uncovers" itself, since without participation it would be absent and because of that cognitively unavailable. Hence every being should be grasped as a necessary relation of one and multiplicity.

We can find a particular realisation of these assumptions in the *Philebus* dialogue. Plato presents here the form of superior knowledge which is available to those who "live closer to the gods" and they are aware that everything that in any way exists consists of a certain one and multiplicity, which come down to limit (*peras*) and indeterminacy (*apeirian*).[5] This way Plato refers directly to Pythagorean tradition and submits it to his own interpretation. He claims that every relation of one and multiplicity, limit and indeterminacy constitutes a certain *compositum*, which in the contemporary language might be legitimately defined with the name of structure. For structure names a certain kind of order or, as Plato says, the way of orderliness (*diakekosmenon*) of the constituent elements. Those elements present multiplicity in the structure, whereas one corresponds to the way of orderliness.

What is one, the source of order in the structure? Plato believes that it should be connected with the notion of limit (limits) which as a form organizes the form of the indeterminate (multiplicity of elements). Limit (limits) determines a kind of structure and the ways to organise multiplicity of elements within its scope.[6] Plato calls limit (limits), ideas. They establish determinacy measures of any multiplicity which is indeterminate itself and demands determining. One can say that ideas (limits-measures) make a kind of "inner schema" marking the form of the specific structure and determining its order.

---

[4] *Ibid.*, 142c–143a.
[5] Plato, *Philebus*, 16cd.
[6] *Ibid.*, 23a–26d.

Ideas are "regularities" which organize the form of structures and the connection principles of any multiplicity (elements) within their scope. Indeterminate multiplicity can be grasped merely in the context of limits which it is entitled to. Plato believes that knowing how many limits there are and what limits they are, only then make man wise and reasonable.[7] If we assume that every structure is the relation of one and multiplicity, limit and indeterminate, then cognition of the structure focuses on the cognition of limits and their kind. It is limits that decide about the form of order in the structure, and what is indeterminate is unknowable. This conviction may be recognized as one of the most important ones which appear in the Platonic conception of knowledge (science).

Plato refers to examples from the field of grammar and music. In his opinion, the language structure makes up a certain indeterminate multiplicity, the multiplicity of alphabet letters. Only to know what letters they are and how they are connected with one another, makes man a language expert and likewise in music. It is an extremely volatile and dynamic structure created from sounds. Its proper cognition is connected with the familiarity of the limits which are present in it and constitutive of it: intervals, rhythms, time and harmony. Only knowledge of them makes man a musical expert. Thus, in the case of language and music, the familiarity of musical ideas, limits, determinacy measures of linguistic and musical matter become decisive in their cognition. Ideas (limits, measures) are not and cannot be dependent there on any subjective determination. It is not a subject that decides whether an octave is an octave, a fourth is a fourth or a fifth is a fifth. Therefore, the cognition of structure becomes identified with the cognition of limit-measures, which decide about the organizational manner of indeterminate multiplicity in being. The organizational manner means order.

What makes this order possible? Is it only the result of a simple superimposition of limits on indeterminacy? Plato thinks that in case

---

[7] *Ibid.*, 17ae.

of structure we deal with a complicated and complex network of connections and relations which occur among elements within its scope. In the *Theaetetus* dialogue, he claims that the examination of a structure cannot be restricted into an examination of its elements but, first of all, it must take into consideration the examination of connections which he calls systems and coincidences of elements.[8] Elements (the elements of structure) are knowable solely in the context of the whole system of references to other elements. Plato says, referring to the example of grammar:

> but the things composed of these are themselves complex, and so their names are complex and form a rational explanation; for the combination of names is the essence of reasoning. Thus the elements are not objects of reason or of knowledge, but only of perception, whereas the combinations of them are objects of knowledge and expression and true opinion.[9]

If we recall that Plato defined knowledge as true belief confirmed by reason, then its essence comes down to reading the structure of connections and relations which occur between elements. Elements themselves are not suitable for a strict depiction, thus they cannot constitute the basis of knowledge. Elements receive their determination only in the perspective of references to other elements. In his *Sophist* dialogue, Plato repeats and confirms his stance. He says that the examination of a certain state of affairs requires a consideration of the ways in which the elements within a certain whole are connected and which of them are possible, which are necessary, or which are totally impossible. He says:

> And certainly one of these three must be true; either all things will mingle with one another, or none will do so, or some will and others

---

[8] Plato, *Theaetetus*, 202ae.
[9] *Ibid.*

will not (…). This is the knowledge and ability to distinguish by classes how individual things can or cannot be associated with one another.[10]

In the *Sophist* dialogue he examines the connections between the greatest kinds (*megista gene*). They are namely being, rest, motion, the same and the different. These connections are not facultative. Their analysis should be carried out to grasp the essence. This analysis is the aim of the dialectical method which examines in what way certain things or state of things connect with one another and in what way they do not.[11] During the analysis appears the distinctions between set and structure. It is exemplified by concepts in sentences. A sentence is not a simple sum of concepts but its sense and essence result from the proper ordering of those concepts (the rules of grammar and logic). On the contrary, a set is treated by Plato as a simple sum of elements which does not possess any determinacy whereas structure is the order of elements, a complex of relations determined by proper measures-limits which give elements determinacy, form and function. Therefore, the task of science is not to examine sets of elements as their simple sum, but to reconstruct the order valid within structures which are created by those elements.[12] In this way, the proper and fundamental subject of science becomes the examination of structures. On this basis we can claim that, in accordance with a contemporary definition, Plato could be numbered among the circle of thinkers who represented the structuralist stance.

The realisation of such an implied stance may be found in the *Timaeus* dialogue. In it Plato makes an attempt to examine the structure of structures which is the Cosmos. He argues that it is created from an indeterminate element, space (*hora*) which does not possess any form, establishing the virtual substratum of the physical nature of

---

[10] Plato, *Sophist*, 252e–253e.
[11] *Ibid.*, 251e–253a.
[12] *Ibid.*, 261d–262d.

Cosmos. This space is unordered. It does not possess any structure or elements. As it is unordered, it is unknowable. The Demiurge, while organizing the order of Cosmos, "superimposed" defined limits and ideas on it. They make the pattern, eternal model, schema, according to which the body of Cosmos was created. An "eternal pattern" makes the indeterminate space determinate and organised. Cosmos originates as the fusion of body and soul, since in Plato's language, the superimposition of the pattern on shapeless space corresponds with the fusion of rationality with corporeality. The physical Cosmos comes into existence.[13]

Ideas, limits, with the use of proportions, create the form of primary elements: earth, water, air, fire and ether. Those elements should not be understood literally. Since they correspond solely with the "states" of the world (today one should say phase states): solid, liquid, gaseous and diverse forms of energy. As proportions and measures may be most precisely expressed and described in the language of mathematics (mainly geometry), Plato became convinced that this very language is predisposed to the description of the structure of the Cosmos. The comprehension of the world is tantamount to its mathematical description. This moment seems decisive in Plato's activity. It influenced the later fate of science and philosophy, because it gave rise to the conception of mathematical natural science (*physiotera ton mathematon*). This conception reached its full development within the walls of Plato's Academy.

I do not regard the discovery of the structural character of Platonic science as original. If under the term of structuralism one understands (in the most general sense) the examination of structures from the position of connections and relations which occur among constituent elements, then Plato's proposition fulfils such conditions. The interpretation of this structuralism appears more difficult. I would like to present the version which seems interesting to me, first of all from the perspective of the contemporary contention about the nature of mathematics.

---

[13] Plato, *Timaeus*, 28a–40b.

Plato often referred to Pythagorean intuitions in his philosophy which he supported ardently. The Pythagoreans were inclined to treat mathematics as the world ontology (later some academicians will follow this belief), arguing that the world is a substance created according to the mathematical schema. Therefore, numbers and geometrical figures are treated in it as primary beings, as beings, *sensu stricto*. The world is mathematically knowable, claimed Pythagoreans, thus from its essence it is mathematical, created from mathematical objects (figures and numbers).[14] Plato knows this view but he is not inclined to accept it. He claims expressly and consistently that mathematics does not create the ontic foundation of the world but it occupies an intermediate position between the world of ideas and the world of spatiotemporal phenomena. Actually, it is not only about the intermediate position but about keeping mathematics independent of ideas and phenomena since neither ideas nor spatiotemporal objects are or can be the object of mathematics.[15] Every real state of the spatiotemporal world is always the connection of ideas and indeterminate matter (*Philebus, Timaeus*) and outside this connection cannot be conceivable. A phenomenon is always the *compositum* of ideas and indeterminate matter (space – *hora*). This Platonic conception was later maintained by Aristotle in his theory of form (*morphe*) and matter (*hyle*). He only rejected the possibility of the independent existence of ideas. Plato, on the other hand, while accepting the difference in the existence ways of ideas and matter (space), acknowledged their ontic independence. Ideas are not the same as space, thus they must be independent of it. However, it does not mean that the structure of the world makes up the strict connection of ideas and matter (space). Such a solution shows that within Plato's philosophy one should unequivocally distinguish the notion of structure from the notion of sys-

---

[14] Philolaus, *Fragment 4*, [in:] *Philolaus of Croton: Pythagorean and Presocratic*, translated by C.A. Huffman, Cambridge University Press, Cambridge 1993; Stobaeus, *Anthology*, I 21, 7b, [in:] *Ioannis Stobaei Anthologium*, C. Wachsmuth, O. Hense, Berolini apud Weidmannos 1884.
[15] Plato, *The State*, 509d–511e, 529de, 533c.

tem. System is purely the set of objects and interactions which occur between them. Objects and their interactions are always entitled to definite, individual properties. The existence of objects and interactions is independent of any subjective determination. Artificial, man--made systems are an exception. Structure then must be grasped as a way in which elements are connected in systems. It not only concerns relations between objects but also the principles which determine these relations. Systems are always organized in some kind of order. They possess the form of element organization typical of them. This form corresponds exactly with the notion of structure. Every system has a structure but not every structure is a system.

Plato assumed that philosophy and science are the examination of systems and structures.[16] It is about the examination of the ways of system organization, thus the forms of their arrangement. Such an examination can be carried out only by a knower of the subject. Only a subject has the ability to reproduce (more or less adequately) the order present in systems. It is possible due to the cognitive activity which means that the intellectual model (modulus – pattern) of system and its structure can take place. This model solely has the status of mental being whose nature is completely dependent on intellect and therefore different from the nature of objects in systems. To distinguish a model from a system, one must assume that the system structure reconstructed in a model consists merely of places and relations which appear within its scope. In systems we deal with objects and interactions while in intellectual models they are replaced by places and relations. They are intellectual images of objects and structures in systems. In his examples, Plato refers to language, musical and phenomenal systems.[17] Elements existing in them have an object-oriented character. They interact with one another in various ways however, these interactions are not facultative. They are determined by limits, measures (ideas) which describe precisely their range and reach.

---

[16] Plato, *Theaetetus*, 201d–203e.
[17] Plato, *Theaetetus, Sophist, Philebus*.

Limits and measures decide about the structure of element connections in systems. This way intellectual models emerge in intellect, both objects in systems as well as measures and regularities, according to which they are organized. How do these models emerge?

Plato argues that images initially account for the result of sensory perception of the world (true belief) which later is subject to intellect activity.[18] First we observe sensual bodies, their forms, movements and behaviours. Yet we notice something more. We notice the common schemata typical of many phenomena, we observe periodicity of natural rhythms, the recurrence of many states and processes, similarity of structures in objects. Briefly speaking, we notice the presence of many patterns (the cycles of day and night, movement of celestial bodies, rhythms, geometrical and temporal patterns) which are in the observed systems.[19] Patterns indicate directly the definite type of order, certain regularities, explicit or implicit. They are present everywhere, both in us and beyond us. Intellect has an evolutionary capability to read those patterns and create their intellectual models. They emerge through abstraction and are the product of activities affirming the constant occurrence of a certain set of features in a certain class of objects and events.[20] Models, as intellectual images of patterns present in systems, have and must have another ontic status than their equivalents in systems. An intellectual model is the sole product of intellect, unlike a pattern in a system. The sphericity pattern in its existence is not dependent on the determination of the subject. It is not the subject that decides whether a sphere is a sphere in various systems. The subject may only (more or less precisely) grasp the pattern of a sphere and build its intellectual model.[21] When it manages to do so, then it acknowledges that it reached the essence of things since to understand the essence of things means to understand the pattern according to which a thing or a state of thing were organised. To under-

---

[18] Plato, *The State*, 509d–511e.
[19] Plato, *Timaeus*, 47b.
[20] Plato, *Letter VII*, 342ad.
[21] *Ibid.*

stand the essence of a circle or a sphere is to understand the pattern according to which a circle or a sphere, were created. On the intellect level it is always the image of a pattern and not a pattern itself. Intellect creates the images of many patterns. In intellect, the position of the object in patterns is taken by places and relations. They represent objects and interactions in systems. Places and relations do not have and cannot have any object-oriented character (which objects of the system are entitled to). Therefore, mathematical objects, as places in structures of models, do not have any metaphysical, spatial-object-oriented nature, and one should not ask about their properties. Plato refers to the example of celestial bodies. We can depict intellectual patterns determining the form of those bodies and patterns bound up with their movement principles. But the images of these patterns in intellect are not any objects, they are also not patterns present in structures. The intellectual model of a pattern is not the pattern according to which an object is created in the system. The intellectual model of a pattern expresses solely the structural properties of a pattern in the system. A pattern in the system defines the essence of an object, decides what it is. It is always an individual pattern, specific, typical of only this object and not another one.

It is different in the case of ideas. The patterns which represent ideas contain the whole class of objects and events. They are the most general measures and regularities which govern the order of the Cosmos. Individual patterns are their forms. Therefore, each individual pattern (the essence of things) must have its cause and foundation in the general regularity and measure. It is a factor determining the cosmic harmony and correlations among all the systems in the cosmic order. The cognition of object patterns in systems is tantamount to reaching ideas-measures which condition those patterns. The cognition of the essence of a specific circle is connected to reaching the superior measure, idea which allows this circle. Let us notice that although there could be many circles and each of them has a specific essence typical of it, the measure of their organisation (as circles) could be only one. It is, says Plato, a circle "in itself" (*kyklos kat auto*). In

Letter VII Plato explains that one can create intellectual images of the patterns present in spatiotemporal objects (e.g. circle). One can name these objects, define them, build their physical and imaginary models, create a system of knowledge about them. However, it is insufficient unless we refer to the ultimate basis, the measure, which conditions the existence and determinacy of those objects. In the case of circle (circles), it is the idea of a circle, which Plato denominates as "the fifth revelation."[22] He also calls it a proto-pattern. Rightly so, because a proto-pattern should be distinguished from a pattern. In the first case it is an idea, in the second case an individual pattern, the essence of an object or state of things in the system.

There is still an answer to the question: what constitutes the basis and condition to build models? Plato recognized that it is possible thanks to the "divine authority" that a subject is entitled to, and which the ancients denominated the capability of abstract thinking (*afairesis*). It is probable that only man possesses it. Therefore, only he is able to create knowledge. The essence of abstract thinking comes down to the ability to draw out permanent and necessary features from the definite, individual state of things (object-oriented and non object-oriented) without which they could not be what they are. These features directly indicate patterns, they are their manifestation. The intellect can separate features from their physical realisation and place on the intellect level while reconstructing the pattern on their basis. Physically we are not able to separate the shape of a planet from its physical form. But we can do that with the use of intellect, recreating in the mind (by means of abstraction) the necessary features of an object which indicate the pattern of its organisation. We create an intellectual model of the pattern of a certain object. Similarly, we can create intellectual models of proto-patterns (ideas). However, they always remain intellectual models which a subject creates in the process of abstract thinking. There are many such models. They do not make up a random set. On the contrary, they are interconnected, and not in

---

[22] *Ibid.*

a facultative way. When on the level of intellect, models of many patterns appear, one should make an attempt to recreate the network of connections which occur among these models. One should recreate the structure of those connections and it is the principal purpose of scientific research. The structure of intellectual models is solely the structure of places and their relations in intellect. One should consider which of them are possible, necessary or totally impossible. It is the task of the dialectical method.[23] It can be carried out only by means of the definite notions and definite language. For examining spatiotemporal structures, in Plato's opinion, the language of mathematics is the most suitable. Therefore, mathematics cannot be understood (as the Pythagoreans wanted) as the primary level of being which determines the form of the world and its nature. It can be understood solely as a way of a subjective description of structures and systems. It is the reason for which it occupies an intermediate place between ideas and phenomena. Mathematics has solely the status of mental being with all the consequences which can result from subjective limitations.[24] This, however, *the subject of what* mathematics describes, is not dependent on intellect or language. It describes the world order, independent of the subject, which exists objectively beyond itself. This order is not the product of intellect, although it is attempted to recreate with the help of it. A subject may only (more or less adequately) reconstruct this order. It may also make mistakes and create simplistic or false models. It can interpret these models in different ways too. Besides, it can use various procedures and introduce the objects constructed artificially (artefacts). It transpires that such an activity is exceptionally effective and gives unexpected results. For example, while trying to reconstruct a system we build its intellectual model, assuming that we deal with a certain network of relations and places situated in some space. Such space may be a model of physical space or ideal space. In the area of mathematics, it is solely

---

[23] Plato, *Sophist*, 253e.
[24] Plato, *The State*, 529de.

mathematical space and in this space we place the objects constructed by us and examine their behaviours. Induction and deduction are made the conditions to carry out analyses. Places in structures represent positions which, in a certain form of ordering, could be taken by objects in systems. We cannot manipulate objects in systems yet we are allowed to operate on places in structures. Then we examine possible ways of their connections. We "throw" the mathematical objects constructed by us into mathematical space and examine how they "behave" in it. We can also examine the space itself, placing various mathematical objects in it and watch how they affect the "behaviour" of the space. It allows us to reconstruct the structure and compare the reconstruction with a specific system.

The method of examining structures proposed by Plato also allows us to answer the question connected with the nature of mathematical objects. Plato assigns them solely the property of being products of intellect (*dianoia*).[25] Thus, they do not constitute the ontic foundation of the world. There is no objective, independent of the subject, world of mathematics. There is merely an orderly world of objects and principles according to which these objects are organised and constructed. This order can be described mathematically. We can say that the world is mathematical only in view of the possibility of description. One should then distinctly separate the *mathematicality* of nature from the capability of its mathematical description. The world is the order which makes the mathematical description possible. But the world is not mathematical in the sense of treating it as a "product" of mathematics.

In the end, one should also ask about the cause which determines the subjective ability to create mathematics Plato, searching for the answer, refers to the tradition of Greek philosophy. He accepts the conviction that man (with inherent cognitive powers) is an integral part of the cosmic order and is entirely subordinate to this order. Man emerged from the cosmic order and is its integral part and form

---

[25] Plato, *The State*, 510ae.

(microcosmos). It also concerns the supreme authority he exercises – intellectual cognition. Its action must be subordinate to all the regularities and rules of the cosmic order and consistent with this order. Otherwise, life in the Cosmos would be impossible. The intellect possesses a natural (evolutionary) ability to read the principles and regularities of the cosmic order. This capability of "reading" Plato calls *anamnesis*.[26] It is identical to the capability of "recollecting" rules and principles of the order according to which it was created and ordered itself. It is the basis of the capability of reconstructing structure of the cosmic order which is accomplished in mathematical language. Mathematics is possible only because the subject has the capability of reconstructing (recollecting) structure of the cosmic order. Therefore, one should not treat the capability of mathematical cognition (as E. Wigner wants) in terms of miracle.[27] The miracle would rather be the lack of such an ability. The Cosmos would become then unknowable and insusceptible to any description. In this respect, one should accept the argument of those who claim that the Cosmos possesses a feature which allows it to be knowable and susceptible to description. Plato thinks that it is what cognition and the essence of science consist in. It is solely "recollecting" that how there is. Since no man determined the order of nature, he also cannot impose his own principles and rules on it while trying to cognize it. He can try, at best, to reconstruct this order and build its more or less adequate model. Perhaps in this very capability the mystery of creating science is concealed.

[26] Plato, *Meno* 81c, *Phaedrus*, 249bd, *Phaedo*, 73b–74a.
[27] E. Wigner, *The Unreasonable Effectiveness of Mathematics in the Natural Sciences*, "Communications in Part and Applied Mathematics" 1960, vol. 13, pp. 1–14.

Wojciech Załuski

Jagiellonian University

Copernicus Center for Interdisciplinary Studies

On the Relevance of Evolutionary Anthropology for Practical Philosophy[1]

1. Philosophy in Science

The term 'philosophy in science', as understood by Michael Heller,[2] embraces a reflection on three types of questions: (a) "the influence of philosophical ideas on the origin and evolution of scientific theories"; (b) "traditional philosophical problems involved in empirical theories"; (c) "philosophical reflection on the assumptions of empirical sciences." Heller adds, however, that the list of three questions does not raise a claim to completeness. Now, in our view, the list could be supplemented with another question (arguably akin to question (b)), namely, the question about the relevance of scientific theories for solving the traditional problems of philosophy. The present essay will deal with a question of this last type. We shall not, however, pursue a general, methodological discussion of the problem of the relevance of science for solving the traditional problems of philosophy, but will rather provide a case-study – an analysis of a more specific problem falling under this question: we shall reflect on the relevance of evo-

[1] The publication of this paper was made possible through the support of a grant "The Limits of Scientific Explanation" from the John Templeton Foundation.
[2] M. Heller, *Jak możliwa jest "filozofia w nauce"?*, [in:] M. Heller, *Szczęście w przestrzeniach Banacha*, Znak, Krakow 1995, p. 20.

lutionary anthropology for the main problems of practical philosophy (whose two main branches are moral and legal philosophy). The claim we shall defend in this essay asserts that evolutionary anthropology is more relevant for legal philosophy than for moral philosophy (two main branches of practical philosophy), i.e., it can be more gainfully invoked when tackling some important problems of legal philosophy than those of moral philosophy.

2. What Is Evolutionary Anthropology?

Let us start our analysis by clarifying what we shall mean exactly by 'evolutionary anthropology'. Briefly, it shall be construed as a non-trivial view of human nature based on Darwinian evolutionary theory. By 'a nontrivial view of human nature' we mean a view that says more specific things about human things than the clichés of the type: 'there is no fundamental gap between human beings and other animals' or '*homo sapiens* came into being as a result of the same evolutionary mechanisms that led to the emergence of non-human animals'. Now, one can argue that such views are provided by various (as they are called) 'evolutionary perspectives' on human nature:[3] evolutionary psychology, behavioural ecology, sociobiology, or gene--culture co-evolutionary theory. The plural noun 'views' is pertinent here, because these four perspectives (especially the three first ones as opposed to gene-culture co-evolutionary theory) provide views of human nature which are divergent in some important points. In the following considerations by 'evolutionary anthropology' we shall understand anthropology built mainly upon evolutionary psychology. The purpose of the following analyses can therefore be stated more precisely as an attempt to answer the question about the relevance of the view of human nature implied by evolutionary psy-

---

[3] K.N. Laland, G.R. Brown, *Sense and Nonsense: Evolutionary Perspectives on Human Behaviour*, Oxford University Press, Oxford 2002.

chology for the important problems of legal and moral philosophy. There are least three arguments for choosing evolutionary psychology as a basis for reconstructing an evolutionary anthropology. First, the differences between evolutionary psychology, sociobiology and behavioural ecology are not really profound: all these perspectives share an important assumption about the crucial role of evolutionary mechanisms in shaping human behavior; in other words, they assume that many patterns of human behaviour are biological adaptations. Thus, as it seems, the evolutionary view of human nature based on evolutionary psychology can be to a large extent reconciled with the view of human nature implied by sociobiology and behavioural ecology.[4] Second, evolutionary psychology seems to be the most developed perspective from among these three perspectives. Third, arguably only the view of human nature implied by these three perspectives can be called 'evolutionary anthropology' in the strict sense, as it implies that many human behavioural patterns have been shaped by natural selection. The view of human nature implied by gene-culture co-evolutionary theory assigns an important role to non-biological factors in shaping our behaviours, so that it may seem dubious whether in can at all be dubbed 'evolutionary' (in the strict biological sense).

Let us now pass on to a short presentation of the main assumption of evolutionary psychology,[5] and thereby of what we have called 'evolutionary anthropology'. Its main assumption says that the human mind is not a blank slate – *tabula rasa* – upon which everything has to be written by environmental and cultural factors and can be written with equal ease: it is composed of built-in psychological dispositions – computational modules – shaped by natural selection which

---

[4] The main difference between evolutionary psychology and sociobiology on one part, and behavioural ecology on the other, is that behavioural ecology more strongly emphasizes the fact that these behavioural patterns may take various forms depending on an the environment in which they manifest themselves.

[5] J. Tooby, L. Cosmides, *Conceptual Foundations of Evolutionary Psychology*, [in:] *The Handbook of Evolutionary Psychology*, ed. D.M. Buss, Jon Wiley & Sons, New Jersey 2005, pp. 5–67.

play a crucial role in shaping our behaviours. Thus, according to evolutionary psychology, the human mind consists of innate computational modules – psychological dispositions – and these modules are evolutionary adaptations, i.e., their presence can be accounted for by the fact that they enabled our ancestors to best cope with the problems they encountered in the ancestral environment[6] and thereby increased on average their chances of survival and reproductive success. What is important in the context of our further considerations is that evolutionary psychology implies (relying on such evolutionary theories as, for example, a theory of kin selection, a theory of reciprocal altruism, as well as on primatology, and evolutionary game theory) that our predispositions to behave morally (i.e., to take altruistic actions) are the products of evolutionary mechanisms. It must be admitted, however, that evolutionary psychology does not make precise what the *contents* of our various natural moral dispositions are. Accordingly, it is compatible with various more specific views. Before we present two main views of this kind, let us define three forms of altruism, namely, kin altruism, reciprocal altruism and pure altruism. In the case of kin altruism, an agent sustains high costs for the good of a relative without expecting the return of these costs in the future. In the case of reciprocal altruism, an agent sustains high costs for the good of an unrelated person expecting the return of these costs in the future. It should be stressed that reciprocal altruism may take not only the form in which the return from the beneficent act is expected from the actual recipient of the benefits, but also a form in which the return from the beneficent act is expected from a different person than the actual recipient of the benefits – from those who witnessed or otherwise were informed about the beneficent act (Richard D. Alexander calls the former form of reciprocal altruism 'direct reciprocity' and the latter form 'indi-

---

[6] These problems concerned in general survival and reproduction, and in particular, for example, finding mates, succeeding in intra-sexual competition, ensuring the certainty of paternity, deterring the adultery of one's sexual partner, detecting cheats in social exchange interactions.

rect reciprocity).[7] In the case of pure altruism, an agent sustains high costs for the good of an unrelated person without expecting the return of these costs in the future. Now, relying on these three forms of altruism, one can distinguish two varieties of the view saying that human beings are pre-disposed to behave morally. The first variety says that human beings are *narrowly altruistic*, that is, they tend to manifest in many circumstances kin altruism and reciprocal altruism. The second variety says that human beings are *genuinely moral*, that is, they tend they tend to manifest in many circumstances kin altruism, reciprocal altruism and pure altruism. Most evolutionary biologists claim that evolution has not endowed us with the tendency to engage in purely altruistic acts: they maintain that such acts are just the maladaptive side-effects of kin altruism and reciprocal altruism. However, some scholars defend the claim that evolution has endowed us with such a tendency. They either argue that such a tendency is a product of sexual selection or that it is a product of group selection. According to the former explanation, the tendency to undertake purely altruistic acts may have evolved, even though it decreased the probability of survival of those who manifested it, if this tendency was for some reason attractive for the opposite sex and thereby increased the probability of mating and reproductive success. According to the latter

---

[7] Cf. the following quotation: "... returns from indirect reciprocity may take at least three major forms: (1) the beneficent individual may later be engaged in profitable reciprocal interactions by individuals who have observed his behaviour in directly reciprocal interactions and judged him to be potentially rewarding interactant (his "reputation" or "status" is enhanced, to his ultimate benefit); (2) the beneficent individual may be rewarded with direct compensation from all or part of the group (such as with money or a medal or social elevation as a hero) which, in turn, increases his likelihood of (and that of his relatives) receiving additional perquisites; or (3) the beneficent individual may be rewarded by simply having the success of the group within which he behaved beneficently contribute to the success of his own descendants and collateral relatives" (R.D. Alexander, *The Biology of Moral Systems*, Aldine de Gruyter, New York 1987, p. 94). The idea of indirect reciprocity was in fact already hinted at by Robert Trivers (R.L. Trivers, *The Evolution of Reciprocal Altruism*, "Quarterly Review of Biology" 1971, vol. 46, pp. 35–57), who called this form of reciprocity 'generalized reciprocity'. However, it was worked out in detail by Richard D. Alexander (R.D. Alexander, *The Biology of Moral Systems...*, op.cit.).

explanation, genetic group selection may have favoured the tendency to undertake purely altruistic acts, the reason being that groups with many individuals endowed with this tendency and thereby inclined to sacrifice themselves for the group to which they belong are likely to fare better than and win over groups with individuals endowed only with the tendency to display kin altruism and reciprocal altruism. In sum, evolutionary anthropology does not say clearly if we are only kin and reciprocal altruists, or also pure altruists.

3. The Meaning of Evolutionary Anthropology for Moral and Legal Philosophy

The basic claim of this essay, namely, that evolutionary anthropology is less relevant for moral philosophy than for legal philosophy, is based on a certain understanding of morality and law. According to this understanding (contrary, for example, to consequentialist ethical doctrines), the object of moral judgments are in the first place motives of agents, and the object of legal judgments are in the first place acts of agents. Consequently, in order to resolve two parallel problems (one related to moral philosophy, the other related to legal philosophy), namely, whether *homo sapiens* is (statistically) *homo moralis*, and whether *homo sapiens* is (statistically) *homo legalis*, we are compelled, respectively, to probe into the questions of whether human beings are disposed to act on moral motives, and of whether human beings are disposed to behave morally (assuming that the law deals with moral acts, we narrow down our considerations to legal rules constituting Hart's minimal content of natural law, i.e., rules which prohibit morally reprehensible acts directed against such basic human goods as life, bodily integrity or property; law understood in this way can be viewed as a device for promoting cooperative behavior). This linguistic convention is concordant with the Stoic-Kantian distinction between morality and legality of acts, the former concerning their 'depth', i.e., the motives standing behind them, the latter concerning their 'surface', i.e., their external form. In the following considerations, by 'moral motives' we shall understand all of the following three types of motive: (a) the Kantian motive, i.e., a desire to discharge an obligation just because it is an obligation; (b) the altruistic motive, i.e., a desire to promote other people's interests; the altruistic motive is understood here as a manifestation of a pro-social virtue (say, benevolence or generosity), i.e., of an enduring disposition to promote other people's interests; (c) the emotional moral motive, i.e., being moved by morally positive emotions (e.g., gratitude, sympathy).

4. Evolutionary Anthropology and Legal Philosophy

We shall first focus on the problem of whether *homo sapiens* is *homo legalis*. As mentioned before, evolutionary anthropology asserts that human beings have been equipped by natural selection with certain pre-dispositions to behave morally (i.e., in an altruistic and cooperative way). It can therefore be said that according to evolutionary anthropology human beings can be justifiably regarded as *homines legales*, i.e., as having predispositions to *act* morally. This claim can be supported by common-sense observations, for example, the observation that a predominant part of the population does not commit crimes, and more than half of all committed crimes are due to notorious criminals. Clearly, such observations can also be accounted for in a different way; one can, for example, maintain that they can be explained by fear of punishment rather than by in-built dispositions to act morally; accordingly, we do not insist that the former explanation is a better one: we only claim that this explanation is provided by evolutionary anthropology.[8] The question of whether human beings are

---

[8] It is worth noting that in fact one can distinguish at least two different understandings of the concept of '*homo legalis*': in one sense (assumed in this paper) *homo legalis* is a person disposed to behave morally owing to her innate cooperative and altruistic predispositions to behave in this way; in another sense, *homo legalis* is a person disposed to behave morally not thanks to her innate cooperative and altruistic predispositions to

(statistically) *homines legales* is, arguably, 'intrinsically' interesting, though, as may be noticed, it is not a specifically legal-philosophical question. This objection, if apt, is apt only *to a certain extent*, because, arguably, the above question is strictly connected with more typically legal-philosophical questions, for example, the question about the origins of law, and the question of the nature of law. The controversies connected with the former revolve around the problem of whether law could have emerged spontaneously, whereas the controversies connected with the latter question are focused on the problem whether the core of law is an 'expression' of human nature or, rather, a tool for counteracting human nature.[9] Let us devote some attention to the latter problem. Evolutionary anthropology strongly suggests that the core of law is an 'expression' of human nature in the sense that human beings are biologically predisposed to act spontaneously in accordance with legal rules constituting this core, i.e., they are endowed with dispositions whose socio-cultural correlates are these rules. Of course, this core is something more than just an expression of these dispositions: it is also a mechanism for reinforcing those – biologically embedded but nonetheless fragile – dispositions. The problem of the nature of law can also be analyzed from the standpoint of evolutionary anthropology at a more concrete level: one can attempt to show that specific legal institutions (e.g., private property, contract or punishment) are deeply embedded in our biological construction. It should also be noted that, beside those legal-philosophical implications, evolutionary anthropology has also practical implications regarding law, as it can be used to assess the potential effectiveness of various legal regulations in realizing the goals for which they have been enacted. It seems worthwhile to devote some attention to this kind of application of evolutionary anthropology in legal analysis. It is clear that the

---

behave in this way (because she lacks thereof), but because of her fear of punishment. Let us repeat: evolutionary anthropology seems to justify the claim that human beings are (statistically) *homines legales* in the first sense.

[9] W. Załuski, *Evolutionary Theory and Legal Philosophy*, Edward Elgar, Cheltenham 2009.

law achieves its goals (whatever they are) indirectly, i.e., by affecting human behaviour. Consequently, if the legislator does not have a robust model of human behaviour, and does not know how people will react to competing legal regulations from among which she has to choose, she will not know how to make this choice or she may choose regulations which, owing to their inadequate fit with human nature, generate high social costs. Now, according to Owen D. Jones[10] evolutionary theory (and, especially, evolutionary psychology) helps us understand if and to what degree given behavioural tendencies are hard-wired into human nature, and thereby to determine what incentives are necessary to restrain these tendencies or to channel them into cooperative behaviour. He has formulated a law, which he calls the 'law of law's leverage': the law is a conceptual tool inspired by evolutionary psychology to assess the comparative effectiveness of legal regulations.[11] The law says, roughly, that decreasing the frequency of behaviours by means of legal sanctions will be the more difficult (i.e., it will require higher sanctions), the more these behaviours tended in ancestral environments to increase the inclusive fitness of those who manifested them. As Jones and Goldsmith put it, the law "can help us predict and explain the general features of the aggregated demand curves for different behaviours, and thereby helps us understand why some behaviours are less easily manipulated by law than are others."[12] Jones notes that even though the law of law's leverage cannot predict demand curves for law-relevant behaviours with precision, nor can it individualize a curve to a single person, it can offer general and useful insights into the ways the law interacts with various behaviours.

---

[10] O.D. Jones, T.H. Goldsmith, *Law and Behavioural Biology*, "Columbia Law Review" 2005, vol. 105, pp. 405–502; O.D. Jones, *Law and Biology: Toward an Integrated model of Human Behavior*, "Journal of Contemporary Legal Issues" 1997, vol. 8, pp. 167–208; O.D. Jones, *Evolutionary Psychology and the Law*, [in:] *The Handbook of Evolutionary Psychology*, ed. D.M. Buss, Jon Wiley & Sons, New Jersey 2005, pp. 953–974.

[11] The name of this law is connected with the assumption that law exists to effect changes in human behaviour, thus being a kind of lever for moving human behaviour that "depends on a behavioural model as a lever depends on a fulcrum" (O.D. Jones, *Law and Biology...*, *op.cit.*, p. 167).

[12] O.D. Jones, T.H. Goldsmith, *Law and Behavioural Biology...*, *op.cit.*, p. 461.

According to this law, for example, the slope of demand curve for adulterous behaviour will be comparatively steep and thus comparatively insensitive to legal sanctions (because, as evolutionary psychology teaches us, human beings are by nature moderately polygamous); the slope of the demand curve for jealous violence against potential rivals and partners will be much steeper for men than for women (because, as evolutionary psychology teaches us, male sexual jealousy is a biological adaptation). We deem it necessary to end the presentation of Jones's views with a sceptical note: Jones's optimism about the usefulness of evolutionary anthropology in legislation is assuredly precocious for at least two reasons; first, at the present level of its development evolutionary anthropology can hardly provide any precise guidance for a legislator, beyond certain banalities of the sort "it is impossible to eliminate by means of law certain undesirable human dispositions, e.g., sexual jealousy, as they are deeply embedded in our nature" or admittedly controversial proposals (e.g., a proposal concerning a special supervision of stepparents, which is based on the evolutionary discovery that stepparents are much more likely, though 'absolutely' still very unlikely, to commit child abuse or infanticide than biological parents); second, and more fundamentally, it is by no means that the basic assumptions of evolutionary anthropology are true.[13]

Let us summarize. The above argumentation was aimed to show that evolutionary anthropology can be relevant for legal philosophy as it firstly enables one to answer the question of whether *homo sapiens* is *homo legalis*, and is thereby helpful in tackling the legal-philosophical problems connected with this question, namely, problems of the origins of law and the nature of law. Secondly, it can be used (though probably not in a very illuminating way at its present level of development) to evaluate the elasticity of human behaviour and thereby to improve the quality of legislation. One may object, however, that this

---

[13] Other critical remarks on this kind of application of evolutionary anthropology in law can be found in B. Leiter, M. Weisberg, *Why Evolutionary Biology Is (So Far) Irrelevant to Legal Regulation*, "Law and Philosophy" 2010, vol. 29, pp. 31–74.

kind of contribution of evolutionary anthropology to legal philosophy is rather modest. We are inclined to agree with this objection: our claim was not that evolutionary anthropology is an indispensable tool for the legal philosopher but, rather, that it can shed light on some important problems of legal philosophy, and it does not shed any light whatsoever on a number of its other important problems. Especially, evolutionary anthropology *cannot* serve as a basis for the construction of some new version of natural law theory: to believe that it can would be, of course, to commit the 'naturalistic fallacy' (there is a parallel problem in moral philosophy – that of building a normative ethics on the basis of evolutionary anthropology; we shall return to this problem in the next section).

5. Evolutionary Anthropology and Moral Philosophy

We shall now pass on to an analysis of the relevance of evolutionary anthropology for moral philosophy, starting with moral psychology – arguably one of its branches. Moral psychology aims at answering, *inter alia*, the question of whether human beings are *homines morales* (i.e., have in-built pre-dispositions to be *motivated* morally) or not.[14] There are no doubts that moral acts exist (in fact, as was mentioned in previous sections, according to evolutionary anthropology, they should be widespread), but there are serious doubts regarding the frequency of moral acts done out of moral motives (it is clear that non-moral motives may effectively mimic moral motives, i.e., issue in the same acts). One can, roughly, distinguish three views of the relations between moral action and moral motivation:

View 1: Moral acts *always* flow from moral motives.

---

[14] If they are not *homines morales*, then they can be non-moral, i.e., have neither in-built pre-dispositions to be motivated morally nor in-built pre-dispositions to be motivated a-morally, or they can be immoral, i.e., have in-built pre-dispositions to be motivated immorally – egoistically or maliciously.

View 2: Moral acts *always* flow from immoral motives. The best known version of *View 2* is psychological egoism, which asserts that each agent's actions can *always* be plausibly construed as motivated, consciously or unconsciously, by her own self-interest. On this view, morality is a mirage: 'externally' moral acts are always prompted by essentially egoistic motives.

View 3: Some moral acts flow from moral motives, some moral acts flow from immoral motives. This view has two essentially different variants: in its optimistic variant it says that most moral acts flow from moral motives, in its pessimistic variant it says that most moral acts flow from immoral motives.

Now, the problem is that evolutionary anthropology does not enable one to decide between these views, as it does not offer *any* theory of human motivation. It is true that some attempts have been made to appeal to evolutionary theory in order to justify *View 2*, but these attempts, in our view, are unsuccessful – they seem to reflect personal views of human nature of its proponents and have no cogent evolutionary justification. Let us analyze two main evolutionary arguments that have been advanced in favour of this view. The first argument says that, since our evolved dispositions serve the interests of our genes, one can say that 'ultimately' all our moral actions are motivated egoistically – they are intended to serve the interests of our genes. This argument is based on a naïve – literal – reading of the selfish 'gene metaphor': it boils down to a fallacious reasoning that since we are our genes, and genes are selfish, we are selfish, even if we seem to be altruistic. The argument is doubly fallacious – materially incorrect and based on equivocation – since we are neither our genes nor are genes selfish in the sense in which we may be selfish. Thus, this argument conflates two different levels of analysis: the level of the origins of our moral dispositions and the level of motivation standing behind our moral acts. The fact that our moral dispositions are biological adaptations (i.e., serve in the long run the 'interests' of our genes) does not *ipso facto* mean that our moral acts

are motivated by our willingness – conscious or unconscious of promoting our genetic interests. The second argument says that psychological egoism, hypocrisy, and various cheating strategies are often
involved in the relations of direct and indirect reciprocal altruism,
and since it is a dominant form of altruism, we are in fact egoists.[15]
In point of fact, reciprocal altruism is indeed often a sophisticated
form of egoism. This is so when a person engages in reciprocal ex-
changes with the goal of promoting *her* own interest. In such a case,
that person's motivation is undeniably egoistic. It is also true that
agents involved in relations of reciprocal altruism often cheat – in
a gross or subtle way[16] – their partners with a view to maximizing
their own benefits from these relations. However, there is nothing in
relations of reciprocal altruism by virtue of which they must predominantly involve egoistic motivation and be tainted with cheating and
hypocrisy: they can be motivated by positive emotions (e.g., gratitude) or by a sense of duty rather than by an expectation of promoting one's own interests. The above critique of the evolutionary arguments in favour of *View 2* does not, of course, mean that this view
is incorrect. In point of fact, the controversy over the plausibility of
this view is not easily decidable because psychological egoism is an
unfalsifiable theory; nevertheless one can adduce good reasons to reject psychological egoism as a plausible theory of human motivation.

---

[15] This view was of human motivation has been defended by many evolutionary psychologists (cf. R.D. Alexander, *The Biology of Moral Systems...*, *op.cit.*; M.T. Ghiselin, *The Economy of Nature and the Evolution of Sex*, University of California Press,
Berkeley 1974; R. Wright, *The Moral Animal. Evolutionary Psychology and Everyday Life*, Abacus, London 1996); the following famous quotation summarize well this
view: "Scratch an altruist and watch a hypocrite bleed (M.T. Ghiselin, *The Economy
of Nature...*, *op.cit.*, p. 247)." This view was developed in greatest detail by Richard
D. Alexander (R.D. Alexander, *The Biology of Moral Systems...*, *op.cit.*) who applied
his theory of indirect reciprocity to demonstrate that even apparently most disinterested
human acts are motivated by an expectation of return benefits.

[16] The distinction between gross and subtle cheating has been introduced by Trivers
(R.L. Trivers, *The Evolution of Reciprocal Altruism...*, *op.cit.*); in the case of gross
cheating, one of the partners does not reciprocate or reciprocates to her partner so little
that her partner's net gain from the relation is negative; in the case of subtle cheating,
both partners reciprocate but one reciprocates less than the other.

Psychological egoism can be criticized on the empirical grounds that people often act in a manner that can hardly be called self-interested: they jeopardize their lives, health, or well-being for the sake of others; and to try to find egoistic motives behind such acts would be a highly artificial strategy. An adherent of psychological egoism can advance the counter-argument that each agent's *prima facie* altruistic actions are, in fact, self-interested in the sense that whatever the agent does, *she is always merely doing what she most wants to do*, i.e. what she is sufficiently motivated to do. However, should this counter-argument be accepted, psychological egoism would cease to be a bold hypothesis about agents' motivation but, rather, would become an unfalsifiable statement deprived of any interesting content, as it would imply that each agent is selfish in the trivial sense that her motives are *hers and not someone else's*, or in other words, that she maximizes *her* utility function, which is obvious given that *she*, and not someone else, is the subject of *her* actions. Thus, this counter-argument renders psychological egoism tautological. More interesting, though arguably also unfalsifiable and overall implausible, versions of psychological egoism were proposed by some moral philosophers, for example, by Hobbes and La Rochefoucauld. Hobbes maintained (anticipating Nietzsche's moral psychology) that most instances of an agent's moral behaviours are motivated by her expectation of the rise of her sense of power. Thus, for example, an agent's charitable behaviour would be motivated by her expectation of delight she will have in demonstrating (to himself and to the world) her superiority over the person towards whom this behaviour is directed. Pity for a person stricken with misfortune, in turn, would stem from an awareness that the same things might happen to us and undermine our sense of power. This hypothesis is interesting yet implausible since it can hardly be maintained that it applies to every moral act. La Rochefoucauld, in turn, expended a lot of effort to show that all human actions, even apparently the most disinterested ones, are propelled by *amour-propre*, i.e., a desire for self-esteem and for esteem. According to La Rochefoucauld, then, the basic spring of our actions is not egoism in the narrow sense, i.e., care for material goods but a different immoral motive – *amour-propre*.[17] His theory is not trivial or uninteresting (quite the contrary: he gives a detailed and illuminating description of the various incarnations of *amour-propre*), but unfalsifiable (if you have sufficient imagination you shall 'discern' *amour-propre* under the cover of all human actions, e.g., you may argue that the peak of *amour-propre* is to strive to suppress in oneself *amour-propre*) and overall implausible. In sum, psychological egoism does not seem a convincing account of human motivation.

Let us pass on to more general remarks. It seems that the problem of human motivation is too complex and multi-faceted to be effectively tackled by rather crude conceptual tools of evolutionary theory. It owes its complexity to the fact that human beings are endowed with a wide gamut of divergent interests that may become involved in their motivation and with the capacity for abstract thinking and self-reflection that additionally enriches this motivation. Thus, for example, human emotions and empathy manifest a much higher level of complexity (e.g. are multi-layered and entangled in various mechanisms of self-deception and 'transmutations'[18]) than is assumed in evolutionary analyses. Accordingly, even if we counterfactually assumed that evolutionary theory shows that natural selection had endowed us only with morally positive emotions (e.g., gratitude and sympathy) as motives of our moral acts, it would not *ipso facto* mean that the motives standing behind moral acts are moral ones. In summary, one may hypothesize that human motivation is one of those moral-philosophical problems that lie, not for fundamental but practical reasons (i.e., because of the enormous complexity of human motivation), beyond the reach of evolutionary anthropology: it does not have the theoretical potential that

---

[17] Though, in some of his maxims, e.g., those about respecting justice or friendship, he seems to treat care for egoism as an element of *amour-propre*; *amour-propre* could therefore be seen as care for self-interest in a more general sense, embracing not only one's material interest but also a less tangible one.

[18] J. Elster, *Alchemies of the Mind. Rationality and the Emotions*, Cambridge University Press, Cambridge 1999.

would enable one to analyze either at a general level (i.e., when trying to say in general whether human beings exhibit moral motivation) or at a concrete level (i.e., when trying to identify a motivation of a concrete individual) the problem of human motivation.[19] To return to the three views on the relations between moral acts and moral motives: it seems that we can only appeal to common-sense psychological wisdom to decide between these views: this wisdom teaches that *View 1* and *View 2* are not plausible and that it is not clear which variant of the *View 3* is more plausible.

The foregoing considerations are not sufficient to substantiate the claim that evolutionary anthropology is of little relevance for moral philosophy. They must be supplemented by arguments for the claim that evolutionary anthropology offers little aid in the analysis of questions belonging to other than moral psychology branches of moral philosophy, namely to normative ethics and metaethics. We shall not deal at greater length with normative ethics, confining ourselves to the simple observation that evolutionary theory cannot be appealed to when dealing with the questions of this branch of moral philosophy without committing the 'naturalistic fallacy'. But some philosophers argue that evolutionary theory can be gainfully used to tackle one of the basic questions of metaethics – the question about the logical status of moral statements. We do not share this opinion and shall now say why. The point of departure for evolutionary metathicists is the assumption that human beings have become endowed by natural selection with dispositions to act morally, and that thereby one can provide a plausible evolutionary genealogy of many of our moral dispositions. So far so good: this is just the general assumption of evolutionary anthropology. But they also claim that results have implications for the metaethical questions. Painting with a broad brush, one can distinguish three general positions on the implications of evolutionary anthropology for metaethics: the first two positions say that

---

[19] The only exception seems to be the claim that kin-altruistic actions (especially, altruistic actions toward one's progeny, is motivated truly altruistically).

evolutionary theory has important implications for metaethics, the last one, which we shall endorse, says that it has no implications whatsoever for metathics. *Position 1* asserts that by demonstrating an evolutionary genealogy of a disposition to take a moral act *P* one thereby provides an argument for the realist character of a moral norm prescribing *P*. The metatethical position implied by this view is moral naturalism. In order to criticize this position, one can appeal to traditional arguments against moral naturalism, especially 'the open question argument'; we shall not be expatiating on this view, because we think that this argument effectively undermines it. *Position 2* asserts that by demonstrating an evolutionary genealogy of a disposition to take a moral act *P* one thereby provides an argument for the anti-realist character of a moral norm prescribing *P*. The argument for this position goes as follows: acting in accordance with this norm, besides what it may be beyond that, proves to be a means for achieving an evolutionary success; in consequence, a moral norm is 'debunked' as subservient to an evolutionary goal of transmitting one's genes to further generations. In other words: the knowledge of the causes which lead us to make moral judgments strips those judgments of their mysterious character: they are determined by our evolved dispositions; it is therefore an illusion to believe that morality exists 'out there', independent of our dispositions. There are two main variants of this position: ethical skepticism and emotivism. *Ethical skepticism*, defended by M. Ruse, assumes that there are no moral facts; morality is a product of biological evolution designed to fulfill a specific function – to ensure the most effective spread of our genes; moral reality only appears to be objective but this is an illusion (though an expedient one, because the belief that there are no moral facts would undermine its efficiency).[20] Thus, moral norms say something about reality but what

---

[20] M. Ruse, *Taking Darwin Seriously: A Naturalistic Approach to Philosophy*, Blackwell, Oxford 1986; M. Ruse, *Evolutionary Ethics: What Can We Learn from the Past*, "Zygon" 1999, vol. 34, pp. 435–451; P. Woolcock, *Ruse's Darwinian Metaethics: A Critique*, "Biology and Philosophy" 1993, vol. 8, pp. 423–439; J.A. Ryan, *Taking the 'Error' Out of Ruse's Error Theory*, "Biology and Philosophy" 1997, vol. 12, pp. 385–397;

they say is false; more precisely, a belief in objective values is built into our ordinary moral thought and language but this belief is false.[21] *Emotivism* or *moral projectivism* assumes that morality is a projection of our evolved dispositions.[22] Certain qualities that appear to be 'in the world' are in fact generated by the nature of the perceiver's mental life. Thus, moral attributes seem to be 'in the world' but they are in fact appearances: they are caused by our emotional activity. Let us now pass on to the criticisms of this view. *View 2* is intended to refer to all moral systems, so it would be plausible as a general account of a metaethical status of *all moral statements* only if all moral systems could be demonstrated to be rooted in our evolved dispositions. But such demonstration is not feasible for the simple reason that there are incompatible moral systems. Thus, this view cannot be correct with regard to moral systems that set requirements that do not have counterparts in our evolved dispositions (for instance, such exacting moral systems as the Christian ethics or the Buddhist ethics).[23] This position *might* therefore be plausible only with regards moral systems that set requirements having counterparts in our evolved dispositions. But it would be plausible with regard to those systems only if the following three conditions were fulfilled: (i) if evolution equipped us with precisely defined moral dispositions, and thereby did not leave much room for their interpretation, so that moral norms would be *exact socio-cultural correlates* of our evolved dispositions and (in a related

---

W.F. Harms, *Adaptation and Moral Realism*, "Biology and Philosophy" 2000, vol. 15, pp. 669–712; R. Joyce, *Darwinian Ethics and Error*, "Biology and Philosophy" 2000, vol. 15, pp. 713–732.

[21] This variant is based on J. Mackie's error theory, which is a form of 'anti-realist cognitivism' (J. Mackie, *Ethics: Inventing Right and Wrong*, Penguin, London 1991). A view which is similar to ethical scepticism is defended by Richard Joyce (R. Joyce, *The Evolution of Morality*, MIT Press, Cambridge, Mass. 2006); he calls this view *moral factionalism*; it consists of four theses: moral judgments are not in fact true; we may come to know that they are not true; they are nevertheless useful; but they are useful only if we treat them as true in our day-to-day lives.

[22] R. Wright, *The Moral Animal. Evolutionary Psychology and Everyday Life*, Abacus, London 1996.

[23] The Christian ethics, for example, prescribes such 'anti-evolutionary' duties as love of one's enemies, not resisting evil, unconditional forgiving etc.

way); (ii) if one could not provide other reasons for following these norms than evolutionary ones; (iii) if demonstrating an evolutionary genealogy of an agent's disposition to take a moral act P were tantamount to demonstrating that an agent's motive to take this act is to foster his or her evolutionary success. *But none of these premises is fulfilled*: our moral dispositions have a very general and imprecise form; one can with the greatest facility adduce non-evolutionary reasons for complying with moral norms; the fact that a given moral disposition has evolutionary origins does not, of course, mean that an agent follows a norm correlated with this disposition because she can find an evolutionary rationale for this norm. Thus, *Position 2* seems entirely unconvincing. *Position 3* asserts that demonstrating an evolutionary genealogy of a disposition to take a moral act P has no implications whatsoever for the question about the logical status of a moral norm prescribing P. This view, given the above critical arguments against the previously discussed views, seems to be most plausible. In sum, the analysis pursued in this section leads to a more general conclusion that the problems of normative ethics and metaethics lie, for fundamental reasons (i.e., because they are entangled in the deeply philosophical issue of one's general outlook of the world), beyond the reach of evolutionary anthropology and the natural sciences in general.

Conclusion

We have defended the claim that evolutionary anthropology has limited relevance for the traditional problems of moral philosophy. It does not help in deciding between various normative moral systems proposed by moral philosophers and between various metaethical views. In other words, after becoming acquainted with evolutionary anthropology we can just as legitimately accept all of the normative moral systems and metaethical views we were inclined to accept beforehand. Moreover, evolutionary anthropology has little to say about the most interesting problem of moral psychology, i.e., the problem of the motivation of our moral actions. It does not mean, of course, that evolutionary anthropology does not contribute anything to moral psychology (e.g., there is very interesting work drawing on evolutionary theory on self-deception or on happiness).[24] Nonetheless, these are not the central problems of moral motivation; and evolutionary anthropology does not offer a clear account of human motivation – the central problem of moral psychology. We have also defended the claim that evolutionary anthropology may have less limited relevance for legal philosophy. This claim was based on the assumption that, while moral philosophy deals above all with human motivation, legal philosophy is focused in the first place on human action. Now, because evolutionary anthropology reveals human behavioural tendencies (even though it sheds little light on motivation standing behind actions that issue from these tendencies), it is highly relevant for a debate about human nature, i.e., the question of whether human beings are *homines legales* or not (clearly, if we assumed than moral philosophy deals with actions rather than motivations, our argument for the claim that evolutionary anthropology is less relevant for moral philosophy than for legal philosophy would become much weaker; nonetheless, we believe that, given that moral judgments are focused, at least in everyday moral discourse, on motives rather than mere actions, the primary object of moral psychology is motivation, not action: moral actions, however paradoxical it may *prima facie* sound, are more interesting for legal philosophy than for moral philosophy). As was argued, this question has far-reaching implications for more typically legal-philosophical questions – about the origins and the nature of law (its conventional or natural character). We have also argued (following Jones, though with much less enthusiasm than he does) that evolutionary anthropology could be useful in improv-

---

[24] Cf. R.L. Trivers, *The Folly of Fools: The Logic of Deceit and Self-Deception in Human Life*, Basic Books, New York 2011; J. Haidt, *The Happiness Hypothesis: Finding Modern Truth in Ancient Wisdom*, Basic Books, New York 2006.

ing the quality of legislation, as it provides insights into the degree of 'elasticity' (i.e. susceptibility to change by means of legal sanctions) of various human behaviours.

We would like to conclude our analyses with the following two remarks. First, our analyses have been hypothetical in the followings sense: we do not assume that evolutionary anthropology as presented above is an apt description of human nature. Such an assumption would be extravagant given the various criticisms of evolutionary psychology (aimed, especially, at its basic tenet that many human behavioural patterns are biological adaptations). Our intention was to reflect on the relevance of evolutionary anthropology for moral and legal philosophy without deciding whether this anthropology is true or false. Second, the problem of the relevance of evolutionary anthropology tackled in this paper inscribes itself into a broader problem of the limits of the naturalization of practical (moral and legal) philosophy. We have tried to show that the limits are twofold: fundamental – in the case of metaethics and normative ethics, and practical – in the case of human motivation.

Łukasz Kurek

Jagiellonian University

Copernicus Center for Interdisciplinary Studies

Emotions from a Neurophilosophical Perspective[1]

Introduction

It is difficult to point to an area within the philosophy of mind which has been influenced by science more than the discussion pertaining to the concept of *emotion*. Arguably, the development of the science of emotion made it the case that to be truly informative, contemporary philosophical accounts of emotion must refer to the experimental data provided by neuroscience and psychology. Despite still being in an immature state of development in comparison with physics, for instance, neuroscience and psychology have already provided significant information pertaining to the emotional life of humans, and other organisms.[2] The use of empirical data in the philosophical analysis of the concept of *emotion* is, therefore, a typical example of science in philosophy, i.e. philosophical analysis informed by science.

---

[1] This paper was written within the research grant "The Limits of Scientific Explanation" sponsored by the John Templeton Foundation.

[2] One could point, for instance, to the groundbreaking work on the facial expression of emotions by Paul Ekman and his collaborators (P. Ekman, W. Friesen, *Unmasking the Face: A Guide to Recognizing Emotions from Facial Expressions*, Malor Books, Cambridge, Mass. 2003), or to the work on the neuroscience of fear by Joseph LeDoux and his collaborators (J. LeDoux, *The Emotional Brain: The Mysterious Underpinnings of Emotional Life*, Simon&Schuster, New York 1998).

However, it is interesting to note that with the development of the science of emotion, certain philosophical issues within this domain have come out into open. The presence of these philosophical issues in the science of emotion derives from the fuzziness of the category of emotion. Both at the conceptual level, and at the level of psychological and neuronal mechanisms underlying emotions, there are different phenomena, often difficult to reconcile, which are categorized as emotions. The philosophical issue in the science of emotion is, therefore, whether emotions form a coherent category at all. This issue boils down to the question about the nature of emotions which philosophers should be, perhaps, especially apt to answer. This problem is especially evident when tackled from the perspective of neurophilosophy, i.e. philosophical analysis informed by the experimental data from neuroscience, and, to some extent, from psychology.

The analysis of the above-mentioned philosophical issue in the science of emotion, from the perspective of neurophilosophy, is the overarching goal of this paper. In the first part of the paper, different levels of the explanation of emotional behaviour are discussed. Here, the discontinuity between the philosophical and scientific account of emotions will be underlined. In the second part of the paper, the discontinuity between these accounts of emotion will be discussed in the context of the familiar distinction between cognitive and non-cognitive theories of emotion. In the third part of the paper, the relation between different levels of the explanation of emotion will be presented from the perspective of the issue whether emotions form a natural kind.

1. Levels of Explanation of Emotions

One of the topics discussed in the domain of neurophilosophy is the issue of the relation between different levels of explanation of human behaviour. In particular, the distinction between different levels of explanation can be utilized in the domain of the explanation of emotions. From a neurophilosophical perspective, emotions can be explained at least on three levels: the neuroscientific, the psychological, and the philosophical. All of these three levels of explanation differ, both in the concepts deployed, and in the scope of their explanation. Neuroscientific explanation can be understood as the most specific. In the domain of neurophilosophy, neuroscientific explanation is understood as the explanation of cognitive and behavioural phenomena with reference to the functioning of neural mechanisms.[3] Psychological explanation is a more general type of explanation. It also points to the mechanical basis of cognitive and behavioural phenomena, however, these mechanisms are of a psychological character. Both of these types of explanations refer, therefore, to a certain kind of mechanism. In the discussed context, mechanism is the view that the behaviour of the agents and their cognitive faculties can be explained in terms of the functioning and the organization of the parts of the agents mind, and/or brains.[4]

The most general level of explanation of cognitive and behavioural phenomena is philosophical explanation. From the three discussed types of explanation, this is the one which is the most different. This kind of explanation utilizes what are usually described as mental concepts (concepts such as *fear, anger,* or *joy*). It is important to note that these concepts are also utilized, to a lesser extent, at the neuroscientific and psychological levels of explanation. However, what is crucial is that at the level of philosophical explanation these concepts are understood as propositional attitudes, i.e. psychological attitudes towards propositions. The peculiarity of these attitudes consists in the ability of these attitudes to track the rules of rationality.[5] At this

---

[3] Of course, neuroscientific explanation pertains to the phenomena at the lower levels, i.e. explanation of the functioning of single neurons. The lower the level, however, the less interesting it becomes for neurophilosophy. As it was mentioned, the latter is interested in the explanation of cognitive, or behavioural phenomena.

[4] C. Craver, *Explaining the Brain: Mechanisms and the Mosaic Unity of Science*, Oxford University Press, New York 2007, pp. 2–8.

[5] The assumption about the rationality of agents, which is a necessary condition of understanding agents at the personal level of explanation, is, obviously, only an ideali-

level, the explanation of the behaviour of the agent boils down to the
rationalization of her behaviour, i.e. assigning her such a set of attitudes (such as beliefs, desires, emotions, etc.) which will make her
as rational as possible. When one undertakes to explain the actions
of other people while having at their disposal only the ascription of
propositional attitudes, without reference to psychological or neural
mechanisms, the only way to proceed is to rationalize these actions.[6]
As one of the major proponents of these view argues:

> The concepts of the propositional attitudes have their proper home
> in explanations of a special sort: explanations in which things are
> made intelligible by being revealed to be, or to approximate to be-
> ing, as they rationally ought to be. This is to be contrasted with
> a style of explanation in which one makes things intelligible by
> representing their coming into being as a particular instance of how
> things generally tend to happen.[7]

As a consequence of this view, one can argue that even if we could
identify the neural or psychological mechanisms 'responsible' for
a given behaviour, it would not prove essential to the explanation at

zation. The proponents of this view realize it, but despite this rationality is understood
on this view as inherent in this kind of explanation (D. Dennett, *The Intentional Stance*,
MIT Press, Cambridge, Mass. 1987, p. 52).

[6] The classical, philosophical distinction indicates two types of explanation of human behaviour: personal, and sub-personal (D. C. Dennett, *Personal and Sub-personal
Levels of Explanation*, [in:] J. Bermudez, *Philosophy of Psychology: Contemporary
Readings*, Routledge, New York 2006, pp. 17–21). The personal type of explanation
can be compared to the philosophical explanation, which contains the irreducible component of rationality. The sub-personal level of explanation explains away the rational
component of behaviour by invoking specific mechanisms (psychological or neural).

[7] J. McDowell, *Functionalism and Anomalous Monism*, [in:] *Action and Events*, eds.
E. Lepore, B. McLaughlin, Blackwell, Oxford 1985, p. 389. Apart from Dennett and
McDowell, other proponents of this understanding of propositional attitude explanation
of behaviour include Donald Davidson (D. Davidson, *Essays on Actions and Events*,
Clarendon Press, Oxford 1980), Jennifer Hornsby (J. Hornsby, *Simplemindedness: in
Defense of Naïve Naturalism in the Philosophy of Mind*, Harvard University Press,
Cambridge, Mass. 1997), or Alan Millar, (A. Millar, *Understanding People. Normativ-
ity and Rationalizing Explanation*, Clarendon Press, Oxford 2004).

the most general, propositional attitude level. The reason for this is that, at his level, the explanation of behaviour consists in offering an account of how people should rationally behave, and not only of why people actually behave as they do.

Perhaps in the case of emotions, rationalizing explanation seems to be less obvious. Intuitively, emotions seem to interfere with rational behaviour. However, on closer inspection, the rationality of emotions comes into full view. The rational component of the explanation of emotion boils down to the fact that they can be correct or incorrect. Their correctness and incorrectness, in turn, is a consequence of the fact that being in an emotional state can be connected with two kinds of mistakes. First of all, the state of affairs which causes the emotion could be misrepresented by the agent (e.g. it is not the case that the person I am angry at committed the act). The rules of rationality which govern this evaluation are, therefore, rules which determine when a representation of the situation eliciting emotion is appropriate. Secondly, despite the state of affairs being represented correctly, the emotion can still be described as incorrect because of the inappropriate standards of evaluation of the state of affairs possessed by the agent (e.g. a person is angry at another person despite the fact that the act committed was insignificant). Rules of rationality of the second kind are rules which determine when the evaluation itself is correct.

Arguably, a fully-fledged neurophilosophical account of emotions should refer not only to the philosophical level of explanation, but also to neuroscientic and psychological levels. Furthermore, any account which undertakes to combine different level explanations of a given phenomenon should utilize not only the components of explanations present at each level (e.g. psychological or neural mechanisms) but also components allowing us bind the different levels of explanation. A useful terminology in this context is proposed by Jose Bermudez.[8] Bermudez described the specific explanations

---

[8] J. Bermudez, *Philosophy of Psychology*, Routledge, London 2005, pp. 32–33.

present at each level of explanation as *horizontal explanations*, whereas the components allowing to bind the different levels of explanation belong to the *vertical explanation* category.

The horizontal explanation of a particular event or state consists in indicating distinct events or states which are, at least usually, temporally antecedent from the events which are being explained. The next stage of this kind of explanation consists in a proposal of a relation between these events. The paradigm case of a horizontal explanation is causal explanation, where the relation holding between the two categories of events is described as causal.[9] For example, if we want to explain the firing of a neuron, we can refer to the events happening at the synaptic junctions between the neurons as the causes of the firing of the neuron. Such an explanation is a typical case of a horizontal explanation. Furthermore, it is argued that a crucial assumption of this kind of explanation is the assumption about the existence of causal laws.[10]

As Bermudez plausibly claims, in some situations it is not enough to propose only a horizontal explanation of a given phenomenon. In such cases, an explanation of the fact that the above-mentioned relations between the events hold is needed. In the case of the firings of the neurons, a very relevant question would be to ask how, in some cases, the events at the synaptic junctions happen. This question can be answered at the lower level than the level of a whole neuron, for example by reference to the activity of neurotransmitters. According to Bermudez, explaining why horizontal relations between distinct events hold is the aim of *vertical explanation*. What is important here is that often when a vertical explanation is proposed, it consists in explaining the relations at the higher level of explanation by reference to the relations at the lower level of explanation. Therefore, combining horizontal and vertical explanation usually leads to a hierarchical explanation of a given phenomenon.

---

[9] *Ibid.*, p. 32.
[10] D. Davidson, *Actions, Reasons, and Causes*, [in:] *idem, Essays on Actions and Events*, Clarendon Press, Oxford 1980, pp. 3–20.

Perhaps the most informative neuroscientific account of combining the horizontal and vertical explanations is David Marr's explanation of vision.[11] Despite the fact that this proposal concerns only one cognitive ability, namely vision, it was deeply influential among not only neuroscientists, but also psychologists, and philosophers. The main reason for this is that this account is perhaps the most elaborate attempt to connect different levels of explanation of a single cognitive phenomenon into a coherent whole.

The main assumption of Marr's account is that vision is a cognitive ability involving information processing. The processing of visual information consists in conversion of the information coming through the retina (the part of the eye responsible for receiving visual information) to another kind of information, e.g. information about the possible dangers in the organism's environment. The level of explanation relevant to identifying the input and the output of the cognitive system of vision is the *computational* level. This is the most general level of explanation, where the description of the function of a given cognitive ability (in this case, vision) is formulated, and it is explained why this function is performed. Only after this function is properly described, can one engage in the explanation at the lower levels. This general description of the cognitive phenomenon enables one to integrate the explanations at lower levels.

The lower level of explanation is the *algorithmic* level. On this level it is crucial to propose a description of how the system accomplishes the task identified at the computational level. It consists in describing the representations the cognitive system employs and describing how the input representations are processed to produce output representations. The lowest level of explanation is the *implementational* level. At this level it is crucial to localize a physical structure in the brain which realizes the processing of the input and the output representations.

---

[11] D. Marr, *Vision: A Computational Investigation into the Human Representation and Processing of Visual Information*, MIT Press, Cambridge, Mass. 2010.

Marr's account of vision is a typical example of utilizing both horizontal and vertical explanations. At each of the three levels of explanation, distinct horizontal levels of explanation are present. However, taken separately, these levels do not form a fully-fledged explanation of how visual information is processed in the brain. Of course, one can argue that the most basic level in this example, the implementational level, is the level where the genuine explanation of vision is located. However, this argument omits the consolidative role of the higher levels of explanation, especially the most general, computational one (it can also be described as functional). Without answering the question as to what is the function of a given cognitive ability, and why the system performs that function, the search for the specific mechanisms realizing this general function will be undirected.

However, one should note that this multilevel explanation concerns only a narrow and highly specialized cognitive ability. Even more so, this explanation deals only with *early visual processing*, which is the visual processing from the moment of receiving the initial information about external environment to the moment of constructing a three-dimensional object from the input information. Therefore, the visual recognition of the perceived object, the remembering of the perceived objects, and many other cognitive abilities related to vision are left out of the scope of the explanation. This remark is crucial because explanations of low-level, specialized cognitive abilities do not seem to involve more general concepts deployed at the higher levels of explanation, and, as a consequence, there are no significant difficulties with proposing a vertical explanation of these abilities.

Low-level cognitive abilities are usually understood as modular cognitive processes. In cognitive science, modules are described as distinct faculties of the mind, which produce quick solutions in highly determinate circumstances. Some of the other typical properties of such modules include:

(1) Mandatory application. Modules respond automatically to stimuli of the appropriate kind, rather than being under any executive control.

(2) Informational encapsulation. Modules are unaffected by the activity of other parts of the mind. They cannot be 'infiltrated' by background knowledge or the expectations of the agent.

(3) Fixed neural architecture. It is often possible to identify determinate regions of the brain associated with particular modules.

Early visual processing seems to possess all of the above-mentioned properties. However, high-level cognitive abilities cannot be plausibly described as modules of this kind. Such cognitive abilities as problem solving, decision making, or reasoning certainly cannot be justifiably characterized as producing quick solution in highly determinate circumstances, being outside executive control, or being unaffected by the activity of other parts of the brain. It is exactly the opposite. These cognitive abilities are flexible, heavily influenced by different kinds of information, and under the cognitive control of the agent. Of course, even at this level, with the help of some generalizations one can propose certain roles which those cognitive abilities perform. However, those roles will not be as specific as it is in the case of modules. One can suspect that their vertical explanation will be similarly coarse-grained.

The explanation at the higher, philosophical level, which refers to propositional attitudes is even more problematic. The deployed concepts are usually vague, and the horizontal explanation of behaviour which refers to *anger*, or *disgust* is, therefore, understandably imprecise. The situation is even worse if we try to propose a vertical explanation of the relations between actions and emotions with reference to psychological and neural mechanisms. The rational component present at the highest level of explanation disappears at the neuroscientific and psychological levels. This underlines the fact that accounts similar to Marr's cannot be utilized, at least straightforwardly, at the level higher than the explanation of simple cognitive functions.

These difficulties become evident when one undertakes to describe what the explanation of behaviour in terms of propositional attitudes consists of. The general description of this type of explanation could point to the fact that on this level the mind is treated as a whole. It is not, therefore, divided into modules. There are several more specific accounts of how to explain the abilities to explain behaviour and cognitive phenomena with reference to propositional attitudes. One of the most popular states that agents possess a theory of behaviour which is represented in their brain. This theory is supposed to be implicit and consists in broad generalizations which indicate how different propositional attitudes cause other propositional attitudes, and, in turn, how propositional attitudes cause behaviour. Such accounts of the discussed ability are often criticized because they offer only crude generalizations and omit the fact that often the abilities which allow to predict and explain behaviour consist in utilizing simple heuristics which do not employ generalizations of this kind.

An argument against such an account could point to a game-theoretic example of a TIT-FOR-TAT strategy. This strategy is composed of only two rules: (1) always cooperate in the first round; (2) in any subsequent round do what your opponent did in the previous round. If agents do employ this simple strategy in predicting and explaining behaviour in some circumstances then the range of explanation in terms of propositional attitudes is restricted because this strategy does not utilize concepts of these attitudes. To apply this strategy, one must only need to identify whether the other player cooperated or defected.

On the other hand, it is sometimes argued that game-theorists essentially employ concepts of propositional attitudes such as *belief* and *desire*. The difference, however, is that on the game theoretic account, these concepts are 'disguised' as the concepts of *probability* and *utility assignments*.[12] Those plausible criticisms are, however, usually limited to a certain kind of situations in which the reference to the un-

---

[12] P. Pettit, *Decision Theory and Folk Psychology*, [in:] *idem, Rules, Reasons, and Norms*, Oxford University Press, New York 2002, pp. 192–221.

conscious employment of quite complex generalizations seems to be an *ad hoc* hypothesis. Despite this fact, these considerations indicate that the gap between higher level explanation of emotion, i.e. explanation in terms of propositional attitudes, can be, at least in certain situations, overcome.

The above discussion pertaining to the relation between different levels of explanation allows for a formulation of the thesis of discontinuity between higher and lower levels of explanation of emotion. This thesis can be formulated in two versions, the weak and the strong. The former version of the discontinuity thesis states that:

> Weak version of the discontinuity thesis: The propositional attitude explanation of emotions concerns higher cognitive functions which cannot be plausibly described as being realized by simple cognitive modules. It is, therefore, not possible, at least at the present moment, to propose a complete neuroscientific or psychological explanation of these functions (in terms of neural or psychological mechanisms).

The weak version of the discontinuity thesis states, therefore, that there is a discontinuity between the different levels of explanation. However, this discontinuity is not radical. The thesis underlines the fact that at the present moment both neuroscience and psychology cannot propose a fully-fledged account of emotions. Some of the functions of emotions are too complex to propose such an account. They involve higher cognitive phenomena which elude modular characteristics.

On the other hand, the strong version of the discontinuity thesis states that:

> At the highest level of explanation of emotion there exist inherent constraints of rationality. This kind of explanation is therefore qualitatively different from the explanation at the lower levels, e.g. at the level of neuroscientific or psychological explanation, because there are no equivalent constraints at these levels of explanation.

As it was already mentioned, the main argument for the strong discontinuity between the propositional attitude level of explanation and the lower levels of explanation (especially the neuroscientific explanation since this kind of explanation sometimes manages without employing mental terms at all, e.g. Marr's account of vision) does not rest on the fact that this explanation is imprecise, or that it concerns higher cognitive functions which cannot be plausibly described as being realized by simple cognitive modules. This argument is grounded in the observation that at the highest level of explanation there exist inherent constraints of rationality. This kind of explanation is, therefore, qualitatively different from explanation at the lower levels.

What is more, one can argue that the propositional attitude explanation should be at the top of the explanation hierarchy because of its consolidative role. It is, therefore, indispensable at least to the explanation of the cognitive abilities of the higher order. As it was the case with Marr's account of vision, without the highest level of explanation, the search for more concrete psychological or neuronal mechanisms underlying behaviour would be undirected.

2. Cognitive Versus Noncognitive Theories of Emotion

There is no doubt among philosophers and scientists who study emotions that they play a significant role in the domain of decision making, forming evaluative judgments, and moral behaviour to name but a few. However, despite this *consensus* about the role of emotion, there is fundamental disagreement about the issue what emotions really are. At the moment there are two distinct and, to a greater extent, conflicting theories of emotions, namely the cognitive theory and the noncognitive theory. Cognitive theories underline that emotions possess content, which means that they are *about* something. For instance, moral emotions such as anger, guilt, or shame are not only feelings, but also judgments about particular states of affair. It follows that emotions are just a distinct kind of a mental state (to be specific, a distinct kind of a propositional attitude), and can be explained at the propositional attitude level. Noncognitive theories, on the other hand, identify emotions with feelings. What is important here is that, at least traditionally, feeling is contrasted with thinking. We can describe this difference in terms of the above-mentioned distinction between low-level and high level cognitive abilities. The former respond in highly specific circumstances, they cannot be 'infiltrated' by background knowledge or the expectations of the agent. This description can also be applied to feelings. The latter do not possess these characteristics. The high level cognitive abilities are usually associated with thinking. According to cognitivists, emotions, at least to some extent, resemble such high level cognitive abilities due to the fact that they involve thoughts.

A good example of a noncognitive theory of emotion is the James--Lange theory. This theory was proposed separately by both William James (in 1884) and Carl Lange (in 1885).[13] According to the James--Lange theory, the commonsense explanation of emotions (and, one could add, propositional attitude explanation as well) is incorrect. According to this kind of explanation, first (1) there is an eliciting event, such as an encounter with a bear; (2) this encounter causes the feeling, which is called fear; (3) the fear causes the body to react and flee. James argued that this sequence of events is mistaken. He famously proposed that emotions do not precede bodily reaction to the eliciting event. When we find ourselves in an emotion eliciting situation, our body reacts automatically and the subsequent emotion, such as fear in the above-mentioned example, is only a perception (feeling) of these bodily reactions.

This general mechanism of how emotions come into being was adapted by neuroscientists. For example, António Damásio defends a similar account of emotion.[14] According to Damásio, however, some

---

[13] W. James, *What is an Emotion?*, "Mind" 1884, vol. 9, no. 34, pp. 188–205.
[14] A. Damasio, *Descartes' Error: Emotion, Reason, and the Human Brain*, Penguin, New York 2005.

are unconscious perceptions of the patterns of bodily changes. Therefore, feeling is not a necessary component of emotion, because we can have emotions without conscious experience (feeling). That is why Damásio's theory is not a classical example of a feeling theory of emotion but can be more accurately described as a somatic theory of emotion.[15]

Another leading neuroscientist in the field of emotions, Joseph LeDoux, proposes a similar explanation of fear:

> The conscious fear that can come with fear conditioning in a human
> is not a cause of the fear response; it is one consequence (and not ob
> ligatory one) of activating the defense system in a brain that also has
> consciousness.[16]

These explanations of emotion are at odds with the commonsense (and propositional attitude) explanation of emotion. According to this type of explanation, emotions cause action. However, on the neuroscientific account, it is emotion which is caused by the perceptions of the prior bodily changes. The mental concept of *emotion* seems to be eliminated from the scope of explanation of behaviour (emotions are epiphenomenal).

Cognitive theories of emotion, on the other hand, are quite similar in their explanation of emotion to the propositional attitude type of explanation. Cognitivists stress that the main component of emotion, which must be present in their explanation, is the thought the emotion involves. It does not mean that feelings (i.e. perceptions of bodily changes) do not play an important part in their explanation. However, a plausible account of emotion should take into consideration the fact that *thoughts* about the object of emotion are integral parts of the emotion itself. Furthermore, these thoughts are of a distinct kind: they evaluate or appraise the object of emotion as important to the

---

[15] *Ibid.*, p. 171.
[16] J. LeDoux, *The Emotional Brain...*, *op.cit.*, p. 147.

agent's well-being (the paradigmatic cases are moral emotions, such as anger, or guilt). For example, anger usually contains a negative evaluation of an act of a different person; guilt contains a negative evaluation of one's act, etc. These appraisals, in turn, cause specific bodily responses which are perceived (felt) by the agent. According to cognitivists, the sequence of events in the emotion eliciting situations is similar to the propositional attitude explanation: (1) perception of an eliciting event; (2) formation of a judgment of the situation; (3) the judgment causes the bodily reactions the perception of which (4) causes the feeling.

In the case of noncognitive theories, there are no essential difficulties with the proposal of the horizontal and vertical explanations of emotion. In the causal sequence leading to the action, the emotion does not play any causal role. Therefore, the horizontal explanation of emotion at the highest, propositional attitude level is eliminated as being false. The correct explanation of emotion should start at the lower-level explanations, where there are no appeals to inherently rational evaluations. As it was already mentioned, noncognitive theories are especially popular among neuroscientists. That is why neuroscientific explanations of emotion usually resemble Marr's hierarchical account of vision which combines horizontal and vertical explanations.

Neuroscientific accounts of emotion such as Damasio's somatic theory, or LeDoux's account are quite similar to Marr's theory of vision because (1) they are comprehensive: these accounts explain emotions at different horizontal levels (at the level of bodily reactions, the level of neuronal activity in the brain, and at the level of conscious or unconscious perceptions of bodily reactions); (2) they offer a vertical explanation of emotion which combines horizontal explanations (the vertical explanation of emotion boils down to the fact that emotions are perceptions of bodily changes); (3) they offer compelling empirical evidence that supports their conclusions (for example, seeing a snake can trigger a fear response before any judgments have time to form because fear can be triggered by neural pathways from the optic nerve to the amygdala bypassing the neocortex which is usually described as being responsible for forming judgment; this evidence obviously undermines cognitive account).

The discontinuity thesis, both in a weak and in a strong version, is, therefore, undermined on the neuroscientific understanding of emotions. On such an understanding, emotions do not possess complex cognitive components, but they are characterised as possessing properties similar to the simple cognitive modules. On this view, each type of emotion is correlated with the functioning of a neural mechanism which is solely 'responsible' for producing emotion of a given type. Despite the fact that a comprehensive, neuroscientific explanation of emotion has been probably proposed only in relation to only a few emotion types (e.g. fear), it seems that there is an implicit assumption that such mechanisms can be proposed to also explain other emotions, especially the basic emotions, which seem to possess distinct biological characteristics (anger, disgust, fear, happiness, sadness, and surprise). On such an understanding, in principle, there are no obstacles in proposing horizontal and vertical explanations of emotion. What is more, on such an understanding of emotion the rational component of explanation is, of course, also made redundant. It is a consequence of the absence of the propositional attitude understanding of emotion.

Despite this, proponents of cognitive theories resist these conclusions because they argue that bodily perceptions lack the rich informational structure that empirical psychology has shown. To demonstrate that emotions involve thoughts, Stanley Schachter and Jerome Singer conducted one of the most famous experiments in the history of psychology.[17] The subjects in this experiment were injected with adrenalin and told that it was a drug designed to improve vision. Some subjects were told that the injection would cause a state of arousal, and others were not. All the subjects were then asked to wait in a room for a short time to let the drug work. In this room, the

---

[17] S. Schachter, J. Singer, *Cognitive, Social, and Physiological Determinants of Emotional State*, "Psychological Review" 1962, vol. 69, no. 5, pp. 379–399.

experimenters planted an actor, who pretended to be a subject in the same study. Some of the subjects were placed in a room with an actor who engaged in silly antics, like throwing a paper airplane and playing with hula hoops. Other subjects were placed in a room with an actor who behaved angrily. Then the subjects were given insulting and funny questionnaires (each subject received one questionnaire). Subjects who had been informed about the effects of the injection did not show any strong emotional responses. However, subjects who had not been informed about it and who were placed with an angry actor and had been given an insulting questionnaire reacted angrily (these subjects did not know why their heartbeat accelerated or why they felt anxious). Subjects who had not been informed about the drug and who were placed with a happy actor and had been given a funny questionnaire reacted happily. The experimenters concluded that although distinct emotions have a common underlying physiological basis, they can be distinguished by the accompanying cognitive appraisal. Furthermore, emotions seem to be clearly influenced by thoughts.

Furthermore, the proponents of cognitive theories often argue that these appraisals are complex, concerning not only simple categories such as loss, danger, or insult.[18] For instance, Klaus Scherer proposes five categories of these appraisals: whether the stimulus is novel or familiar, whether it is pleasant, whether it is relevant to our goals, what is the potential for coping with it, and whether it is consistent with our standards (he calls those appraisals 'stimulus evaluation checks'). Such accounts of emotion only underline their cognitive components.

The weak thesis about the discontinuity between different levels of explanation is evidently supported on the cognitive account of emotions. Apart from noncognitive components, such as the neuronal mechanisms which produce bodily changes, emotions are understood on this account to involve thoughts which are complex

---

[18] K. Scherer, *Profiles of Emotion-antecedent Appraisals: Testing Theoretical Predictions Across Cultures*, "Emotion and Cognition" 1997, vol. 11, no. 2, pp. 113–150.

cognitive phenomena which cannot be plausible characterised, as it was already mentioned, as effects of the functioning of simple cognitive modules. On this view, the vertical explanation of emotions encounters significant difficulties to connect the level of neuronal mechanisms with the level of complex cognitive mechanisms. At least at the present moment, too little is known about the brain to propose how its functioning is related to producing thoughts. However, it is plausible to state that thinking is correlated with the evolutionary 'younger' parts of the brain such as neocortex whereas, according to the neuroscientific account, emotions are usually correlated with the evolutionary 'older' parts of the brain, such as the amygdala. If at least some emotions irreducibly involve thoughts then neural mechanisms 'responsible' for producing emotions are much more complex than it is assumed by the proponents of noncognitive theory of emotions. In this case emotions involve not only neural mechanisms of a more modular character traditionally correlated with emotions (e.g. the limbic system) but also the complex neural mechanisms correlated with thinking.

Furthermore, also the strong version of the thesis about the discontinuity between different levels explanation of emotion can be justified on the cognitive account of emotion. This position is usually endorsed by philosophers who explain emotions with reference to propositional attitudes. These attitudes can be understood as involving thoughts because they necessarily involve evaluations of the emotion eliciting situations. The explanation of emotions in terms of such evaluations underlines their susceptibility to rationalization because on this position emotions should be formed in accordance with the rules of rationality. The conclusion whether emotions are formed in accordance with these rules can be understood, therefore, as an irreducible component of the explanation of emotion.

3. Emotions and Natural Kinds

The above-mentioned considerations have led some philosophers to question the fact that the phenomena which fall under the category of 'emotion' are sufficiently similar to one another to allow for a unified science of emotion.[19] It seems that there is a discontinuity between the neuroscientific accounts of emotions (which all seem to be versions of the James-Lange theory) and psychological, or philosophical accounts of emotion. Whether this discontinuity is understood on a weak or on a strong version, both of them lead to difficulties in proposing a unified account of emotion (containing both horizontal and vertical explanations). Despite their comprehensiveness, the neuroscientific accounts of emotion usually concern the basic emotions, such as fear. This allows for an accurate functional description and explanation of how these emotions are realized in the brain and the body of the agent, however, other more complex emotions are left from the scope of explanation. On the other hand, psychological and philosophical accounts stress the fact that even in the context of basic emotions such a position does not allow for their plausible explanation because they do not take into consideration the thoughts which emotions involve. This is even more evident in the context of non-basic emotions.

The statement that emotions do not form a homogenous category is certainly not new. For instance, this proposition was extensively discussed by Gilbert Ryle.[20] Recently, however, mainly due to the fact of empirical findings from different areas of science of emotion, it has been given a strong justification. The most comprehensive defence of the thesis that emotions do not form a natural kind was proposed by Paul Griffiths.[21]

---

[19] P. Griffiths, *Is Emotion a Natural Kind?*, [in:] R. Solomon, *Philosophers on Emotion*, Oxford University Press, Oxford and New York 2004, pp. 233–249.

[20] G. Ryle, *The Concept of Mind: 60th Anniversary Edition*, Routledge, London 2009, pp. 69–99.

[21] P. Griffiths, *What Emotions Really Are: The Problem of Psychological Categories*, The University of Chicago Press, Chicago 1997. The consensus between philosophers pertaining to this thesis seems to be quite broad. Thinkers who support this position

To propose that a certain category of phenomena forms a natural kind is to state that the phenomena which belong to this category possess natural properties, independent from human understanding of these phenomena. Natural kinds are, therefore, formed by nature, not conventionally by humans who study these phenomena. It is important to indicate, however, how the term 'natural kind' is understood by Griffiths. It is important due to the fact that natural kinds can be understood very differently, and on some formulations the plausibility of there even being such things as 'natural kinds' can be easily impaired. Here is Griffith's explanation of what is understood by him as a 'natural kind':

> The traditional requirement that natural kinds be the subjects of spatiotemporally universal and exceptionless laws of nature would leave few natural kinds in the biological and social sciences, where generalizations are often exception-ridden or only locally valid. Fortunately, it is easy to generalize the concept of a law of nature to the notion that statements are to varying degrees 'lawlike' (have counterfactual force). This allows a broader definition of a natural kind. A category is (minimally) natural if it would be reasonable to place some degree of reliance on some inductive predictions about unobserved instances. This, of course, is a very weak condition. Very many ways of classifying the world are minimally natural. The aim is to find categories that allow reliable predictions in a large domain of properties. The classic examples of natural kinds, chemical elements and biological species, meet these desiderata.[22]

Induction and scientific explanation utilize natural kinds because they allow one to 'project' the observable correlations between the prop-

---

include Aaron Ben-Ze'ev (A. Ben-Ze'ev, *The Subtlety of Emotions*, MIT Press, Cambridge, Mass. 2000), Ronald de Sousa (R. de Sousa, *The Rationality of Emotion*, MIT Press, Cambridge, Mass. 1987), Jon Elster (J. Elster, *Strong Feelings*, MIT Press, Cambridge, Mass. 1999), Robert Solomon (R. Solomon, *Some Notes on Emotion, 'East and West'*, "Philosophy East and West" 1995, vol. 45, no. 2, pp. 171–202).
[22] P. Griffiths, *Emotions as Natural and Normative Kinds*, "Philosophy of Science" 2004, vol. 71, p. 905.

erties of the investigated objects to other objects which belong to the same natural kind.[23] Due to the fact that, for the purpose of scientific research, objects can be categorized into natural kinds it is not necessary to inspect each object belonging to a specific category. One can make discoveries about the properties of the whole category by studying only a small number of members belonging to a given category. Natural kinds are, therefore, important for scientific research because they allow to make reliable predictions in a large domain of properties without the need to inspect every member of the category.[24] According to Griffiths, a typical example of a natural kind is the category of species. It classifies particular organisms according to their morphological, physiological, and behavioural properties. As a consequence, studying only a few members of a given species allows the discovery of the new properties of all individuals belonging to this category.

What is more, natural kinds are relative to the domains of their use. For instance, there are natural kinds which are used by physical theories which do not appear in psychological theories (e.g. the category of electron) and *vice versa* (e.g. the category of fear). What is crucial in Griffiths' proposal is that he argues that emotions do not form natural kinds in the domain of science of emotions. It is not the case that, according to this proposal, we should completely give up using this category. It still plays an obviously significant role in the philosophical discussion about human mind and human behaviour. As Griffiths argues "the claim that emotion is not a natural kind is entirely consistent with the fact that a systematic account can be given of existing concept of *emotion*."[25] However, on this view, the category of emotion does not allow for "a reliable prediction on a large domain of properties" which is the domain of science.

---

[23] N. Goodman, *Fact, Fiction, and Forecast*, Harvard University Press, Cambridge, Mass. 1983.

[24] P. Griffiths, *Emotions as...*, *op.cit.*, p. 905.

[25] P. Griffiths, *Emotion is* still *not a Natural Kind*, http://www.philosophy.dept.shef. ac.uk/AHRB-Project/Papers/GriffithsPaper.pdf, p. 2.

Griffiths argues that emotions fall into at least three different sub-categories. Some are affect programs, which are modular, automatic patterns of response and which have analogues in many other species. Affect programs (the term is borrowed from Paul Ekman) are essentially these emotions which are studied and explained by the above-mentioned neuroscientific theories. The important consequence of this categorization is that affect programs are noncognitive. Emotions which belong to this category are the already mentioned basic emotions. These emotional responses possess characteristics similar to the characteristics of mental modules. They are triggered automatically when the agent finds herself in an emotional eliciting situation, they are brief, and realized by the evolutionary 'older' parts of the brain which humans share with many other vertebrates.[26] As a consequence, they can be found in all human cultures, being biologically hard-wired into the human brain. Only the emotions which can be plausibly characterised to possess these universal, biological properties belong to the natural kind described as an 'affective program'.

The second sub-category of emotions consists in higher cognitive emotions, which are non-modular and cognitively mediated.[27] Griffiths argues that some of these emotions are typical of hominids (e.g. guilt, or jealousy). The crucial difference between 'higher cognitive emotions' and 'affective programs' boils down to the fact that they are realized by different neural mechanisms. Because 'higher cognitive emotions' involve complex evaluations (thoughts) they are realized both by the evolutionary ancient parts of the brain and evolutionary 'new' parts of the brain, typical of hominids.

The last sub-category of emotions are the socially constructed emotions which are non-modular, cognitively mediated and differ among distinct cultures. This type of emotion is also a 'higher cognitive emotion'. The *differentia specifica* of this type of emotions

---

[26] P. Griffiths, *Emotion is* still..., *op.cit.*, p. 4.
[27] *Ibid.*

is, however, the fact that it involves "internalized cultural model of appropriate behaviour."[28] There are fascinating examples from the psychology of emotions which are characteristic to only some cultures but not to others. An example of emotion from Japanese culture which seems to lack a counterpart in Western cultures is the emotion of *amae*. It is the propensity to "depend or presume upon another's love."[29] Another socially constructed emotion seems to be *fago*, experienced by the inhabitants of the Pacific island of Ifaluk which can be described as a composition of compassion, love, and sadness.[30]

Despite the plausible critique of the emotion with natural kinds, the assumption that emotions form natural kinds is evident both in the noncognitive and cognitive theories of emotions. These theories look for a comprehensive, unified account of emotion which could encompass not only different levels of explanation of emotions but also different types of emotions. The difficulties connected with including 'higher cognitive emotions' into noncognitive theories on the one hand, and 'affect programs' into cognitive theories on the other are, however, reflected in Griffiths' distinction between the three, above--mentioned types of emotions.[31] The phenomena belonging to the respective categories possess many properties which prevent their uniform categorization. These difficulties disappear, however, when one accepts the thesis that emotions do not form a natural kind. On this position, both theories of emotions plausibly describe only a part of category of phenomena which they undertake to explain. Whereas noncognitive theories plausibly explain 'affective programs', cognitive theories propose a convincing account of 'higher cognitive

---

[28] *Ibid.*

[29] J. Robinson, *Emotion. Biological Fact or Social Construction?*, [in:] *Thinking About Feeling. Contemporary Philosophers on Emotions*, ed. R. Solomon, Oxford University Press, Oxford and New York 2004, p. 39.

[30] *Ibid.*

[31] To be precise, on Griffiths' account, only the 'affective programs' can be plausibly described as emotions. Other types phenomena traditionally described as emotions differ too much from the phenomena belonging to this category to be characterised as 'emotions'.

emotions'. However, on this position, it should be underlined that it is no longer appropriate to characterise these theories as theories of emotions. These theories become theories of quite distinct types of phenomena which possess different characteristics and which are realised by different mechanisms.

What is even more interesting is the assumption that emotions form a natural kind seems to be a working hypothesis within the *science* of emotion. As the psychologist Lisa Barrett put it:

> Despite the differences in their (emotions – Ł.K.) surface features, many of the most prominent models (of emotions – Ł.K.) share a common set of beliefs about the nature of emotion: Emotions are categories with firm boundaries that can be observed in nature (meaning in the brain or body) and are therefore recognized, not constructed, by the human mind. As a natural kind of emotion, anger, for example, is assumed to be a package of behavioural and physiological changes that are produced by some causal mechanism (in the brain or the mind, again depending on the level of analysis) that is released under certain conditions. Researchers assume that they will know an instance of anger when they see it in the face, voice, or body of another person, or feel it in themselves.[32]

This paradigm in the science of emotion has proven to be very fruitful in the past. It suffices to mention the discovery of several basic emotions which possess universal biological features among humans and other primates as an example of this paradigm. However, on closer inspection, this working hypothesis becomes less evident.

On the basis of the hypothesis that emotions form a natural kind, Barrett forms two empirically testable hypotheses to which there is a significant amount of undermining evidence.

---

[32] L. Barrett, *Are Emotions Natural Kinds?*, "Perspective on Psychological Science" 2006, vol. 1, no. 1, p. 32.

Hypothesis 1: If certain categories of emotion are natural kinds, characterized by projectable property clusters, then it should be possible to characterize each kind of emotion in terms of a suite of distinctive, observable responses that are coordinated in time and correlated in intensity.[33]

As Barrett goes on to observe, in most of the experiments the actual correlations found between experiential, behavioural and physiological measures of emotion proved to be a lot weaker than expected.[34] Despite this discrepancy, psychologists and neuroscientists carry out their research on emotions as if each emotion possessed distinct kinds of characteristics described in terms of subjective experience, facial behaviour, neuronal pattern of activation, and so on. One can even point to typical strategies of explaining the lack of correlation (for instance, social factors masking responses which would otherwise materialize; inability to produce in laboratory studies strong enough stimuli to elicit prototypical emotions; or that the emotional responses vary across individuals).[35]

The second hypothesis derived by Barrett from the assumption that emotions form a natural kind states the following:

Hypothesis 2: Different emotion kinds have distinct causal mechanisms.[36]

Also in relation to this hypothesis there is a large body of evidence which undermines the relation between emotions and psychological or the neural mechanisms which are connected with them. Barrett discusses four types of mechanisms which are commonly related to emotions: mechanisms producing subjective experience, mechanisms producing facial and vocal signals, mechanisms producing

---

[33] *Ibid.*, p. 33.
[34] *Ibid.*
[35] *Ibid.*
[36] *Ibid*, p. 34.

behaviour, and neural mechanisms observed in neuroimaging studies.[37] From these considerations, it becomes evident that the correlations between distinct types of mechanisms and distinct types of emotion is a lot weaker than one could expect on the basis of the assumption that emotions form a natural kind.

Conclusion

If the term 'emotion' does not denote a natural kind then it has some important implications. First of all, there are implications concerning the empirical research on emotion. Experimenters have to be careful in proposing relations between different categories of this phenomena. It would be a mistake to 'project' the observable properties of emotions belonging to one category to emotions belonging to another category. Secondly, it is not possible to formulate a general theory of emotions (noncognitive theory would be more adequate to explain affect programs, cognitive theory would be more adequate to explain higher cognitive emotions and socially constructed emotions). The philosophical assumption of the two different scientific theories of emotion (cognitive and noncognitive) that emotions form a natural kind, which has its origin in the commonsense explanation of emotion, is simply mistaken. It seems that when identified with feelings of bodily changes, some emotions (belonging to the category of affect programs), contrary to this explanation, do not cause behaviour and do not involve evaluations (or even if they do involve the latter, they are just epiphenomena). Therefore, the explanation of these emotions does not involve a reference to the rules of rationality, or to complex cognitive phenomena. In such cases, it does not seem that there is any discontinuity in the explanation of emotion at the different levels (the level of the whole body reactions, the level of neuronal activity in the brain, the level of conscious or unconscious perceptions of

---

[37] *Ibid.*, pp. 34–45.

bodily reactions). The vertical explanation of an affect program can be grounded in a 'perception of the bodily changes' approach. However, other emotions which involve evaluations are still susceptible to the rules of rationality, or are at least connected in some way with complex cognitive phenomena, because of the thoughts they involve. Therefore, there seems to exist, at least in these cases, a discontinuity between the high-level explanation, and low-level explanation of emotion. These difficulties can be at least to some extent avoided, however, if one decides to abandon the position that emotions form a natural kind.

Teresa Obolevitch
The Pontifical University of John Paul II
Copernicus Center for Interdisciplinary Studies

The Issue of Knowledge and Faith in the Russian Academic Milieu from the 19th to the 21st Century[1]

Introduction

In this article, I analyze the relationship between knowledge and faith in the Russian ecclesiastical tradition. The first part contains an outline of the history of academic philosophy. In the second part, I present the development "natural apologetics" in the 19th century and the issue of the relationship between science and religion in the 20th century in orthodox Russia and discuss some methodological aspects of this problem.

The subject of the relationship between knowledge and faith has been raised in Eastern Christian thought from the Early Church Fathers onwards. Over the centuries, the issue has assumed various forms every now and then. Initially (in early Christian apologists) the question was mainly the relation of "pagan" philosophy to the Revelation (Tertullian's famous *Athens and Jerusalem*), afterwards (in the so-called golden age of patristics marked by the teachings of, for example, the Cappadocian Fathers) the philosophers deliberated on the role of the rational inquiries over the fundamentals of faith, which resulted in the development of theology, and since modern times

---
[1] The publication of this paper was made possible through the support of a grant "The Limits of Scientific Explanation" from the John Templeton Foundation.

thinkers have pondered the issue of the relationship between science and faith or between scientific and theological knowledge. All the aforementioned aspects of the subject generally identified as "knowledge versus faith" were examined by Russian theologians and philosophers as well – especially by the lecturers of the ecclesiastical academies and seminaries from the 19th century until the present time.

1. Revelation and Philosophy

Russian universities, somewhat differently to Western ones, did not have theological faculties. The role of the latter was fulfilled by ecclesiastical academies (existing in Moscow, Saint Petersburg, Kiev and Kazan) which prepared the elites of the Russian Orthodox Church (seminaries, being the equivalent of *gymnasia*, were the earlier educational stage). At the beginning,[2] the ecclesiastical academies (the first of which was founded in 1804) the attitude to philosophy was distrustful and sceptical. In accordance with the 1814 act, 10 hours a week were granted for lectures in philosophy.[3] Nevertheless – according to the evidence given by one 19th century historian – it was understood as a discipline whose aim was "experiencing the weakness and helplessness of the human reason in searching for the truth unaided, without the light of Revelation given from above."[4] This is

---

[2] Additionally, it should be mentioned that in the 17th and 18th centuries at higher education establishments – the Kiev-Mohyla Academy in Kiev and Slavic Greek Latin Academy in Moscow (where philosophy was taught since 1634 and 1687 respectively) the language of tuition was Latin, and western Thomistic coursebooks were used in order to prepare the future clergymen to fight Catholicism and the Union of Brest. Later, the scholastic texts were replaced with the coursebooks written in the Protestant spirit and based on Wolff's thought.

[3] See В. Заев, *Реформы духовных академий в XIX – начале XX в. I. Первая реформа духовных академий 1808–1814 гг.*, "Труди Київської Духовної Академії" 2008, vol. 8, pp. 279–282, 287, 297–298, 307–308.

[4] Cit. after: Г. Шпет, *Очерк развития русской философии*, [in:] *Очерки истории русской философии*, eds. Б.В. Емельянов, К.Н. Любутина, Издательство Уральского университета, Свердловск 1991, p. 371.

why the only presented viewpoints were the ones consistent with the
"true reason" of the Holy Scripture. At the same time, Plato was set
as a model – as "the main pillar of philosophy," with the stipulation
that "he should be studied from the original sources, since the philosopher's thought has been distorted by the hermeneutists."[5] Philosophy
was perceived merely as *ancilla theologiae*; metaphysics, history of
philosophy, psychology, logics and ethics taught at the academies were
supposed only to "complement the theological vision of the world."[6]
It is characteristic that since 1817 the issues of religious and secular
education lay within the competence of one dicastery – The Ministry
of Spiritual Issues and National Education, one of the aims of which
was "establishing in the Russian society the salutary harmony between
faith, knowledge and authority (*sic*! – T.O.), or, in other words, between the Christian piety, the enlightenment of the minds and citizens'
existence."[7] As a result, not only the ecclesiastical academies stayed
aloof from philosophy, but also the secular universities as well.[8]

---

[5] Л.Е. Шапошников, *Православие и философия: границы взаимодействия*, "Вече.
Альманах русской философии и культуры" 2002, vol. 13, p. 45.

[6] Cf. Н.А. Куценко, *Профессиональная философия в России первой половины –
середины XIX века: процесс становления и виднейшие представители*, ИФ РАН,
Москва 2008, p. 33; Б.В. Емельянов, *Цензурная судьба русской философии первой
половины XIX века*, "Известия Уральского государственного университета" 2010,
vol. 1 (73), pp. 101–102; Н.К. Гаврюшин, *Русское богословие. Очерки и портреты*,
Нижегородская духовная семинария, Нижний Новгород 2011, pp. 21–23.

[7] Cit. after: Г. Шпет, *Очерк развития русской философии, op.cit.*, p. 443.

[8] Such a subordinationist approach to secular sciences brought about serious detriment
to Kazan University, whose "reformer," M.A. Magnitsky called philosophy the source
of disbelief and heresy, invoking the Epistle of Paul to the Colossians (2:8) where it was
defined as "vain deceit." In the university edifice there was an inscription stating the
misery of the human reason in the face of faith. The outlines of the lectures were subject
to censorship. Thus – as Gustav Shpet states – "some of the professors began to give
lectures in their subjects in an accusatory tone, while others searched for confirmation
of the fundamentals of the Holy Scripture in them. The professor of mathematics found
the manifestation of Divine wisdom in a right-angled triangle, the professor of anatomy
– in the build of the human body. Certain sciences ceased to be taught, e.g. geology,
because all its theories stood in contradiction with the Holy Scripture." (Г. Шпет, *Очерк
развития русской философии, op.cit.*, pp. 451–452). The reputation of Kazan University was later restored by Nikolai Lobachevsky who developed non-Euclidean geometry.
Another wave of deprivation of philosophy at the universities occurred at the times of
Minister Shirinsky-Shikhmatov, who used to say: "The benefit of philosophy has not

Gradually, more comprehensive coursebooks finally appeared, and their authors – the lecturers – initiated the development of the original, "professional" philosophical thought in Russia, the so-called academic philosophy which anticipated the oncoming movement of the slavophiles and university philosophy. The professors of the Academies ceased using Latin and began to teach in Russian. The initiator of the Russianisation in education was, among others, the professor of the Academy in Kiev Ivan Skvortsov (1795–1863), who wrote:

> The Academy needs philosophy in its entirety. It is the necessity of time, and without it the tutor of the Church will not have reverence among his disciples.[9]

Another noteworthy person was the professor of the Saint Petersburg Theological Academy, Fyodor Sidonsky (1805–1873), whose *Introduction to Philosophy* (1833) was considered to be the most outstanding book on philosophy of the first decades of the 19th century. Sidonsky boldly promoted the view about the autonomy of philosophy and its independence from any authority figures. Simultaneously, he also expressed the opinion that philosophical cognition is less complete than theological cognition, although theology requires rational reflection. According to him, "Faith is necessary for the reason by assisting it, and reason is necessary for the faith to develop it and it makes lucid our human awareness of the Divine sphere."[10] Sidonsky's successor in the philosophy department was Vasily Karpov (1798–1867), the translator of Plato into Russian and the advocate of the so-called

---

been proven, while the detriment caused by it is quite possible." In 1850, on the command of the minister, the departments of philosophy at the universities were closed. However, at the time the role of the philosophical centres (facilitating, among others, the reception of German idealism) was taken over by the ecclesiastical academies.

[9] Cit. after: Н.А. Куценко, *Профессиональная философия в России...*, *op.cit.*, pp. 91–92. Cf. also: *idem, Протоиерей Иоанн Скворцов и Киевская духовно-академическая школа*, [in:] *Философия религии: альманах 2006–2007*, ed. В.К. Шохин, Наука, Москва 2007, pp. 393–398.

[10] Ф.Ф. Сидонский, *Введение в науку философии*. Cit. after: Л. Е. Шапошников, *Православие и философия: границы взаимодействия*, *op.cit.*, pp. 48–49.

transcendental synthesis – the postulate of expressing the whole reality, both the empirical and the supernatural one, in a single universal philosophical system (this ideal would be realised later by, among others, Vladimir Solovyov in his doctrine of integral knowledge). The exponents of academic philosophy expressed views which should be considered theistic. In their reasoning, instead of the religious notion of God, they often employed its philosophical equivalents, such as "the absolute", "the absolute being", "the unconditional being", "the infinite being" etc., describing the aspect of God that can be the object of rational reflection[11] and thus retaining the moment of inexpressibility, apophasis, in Him. Philosophy itself was then understood as "the study of being in its relation to what is unconditional."[12]

In the academic milieu, philosophy as *ancilla theologiae* performed the function of the apology of the revealed truth, of "the justification of the forefathers' faith"[13] and was employed to engage with atheistic or agnostic thought.[14] For instance, the professor of dogmatics at the Moscow Ecclesiastical Academy, Filaret Gumilevskij (1805–1866) taught at his lectures that the aim of philosophy is

> to demonstrate how a mystery of Revelation, although it cannot be approached on the principles of reason, does not contradict its theoretical

---

[11] Cf. И.В. Цвык, *Проблема истины в русской духовно-академической философии*, "Вестник Московского университета. Серия 7: Философия" 2004, vol. 2, pp. 14–15.

[12] Г.Д. Панков, *Апологетика философии в контексте апологетики богооткровенной веры в православно-академической мысли*, in: *Колізії синтезу філософії і релігії в історії вітчизняної філософії (до 180-річчя Памфіла Юркевича та 130-річчя Семена Франка)*, eds. Г. Аляєв et al., АСМІ, Полтава 2007, p. 24.

[13] V. Solovyov will return to this idea later.

[14] However, voices criticizing philosophy even as the servant of theology appeared as late as in the second half of the 19th century. For instance, in 1884 the journal "Вера и разум" (*Faith and reason*) issued by Kharkiv diocese, published a text whose author protested against the positive evaluation of philosophy carried out by the professor of the Moscow ecclesiastical academy, Victor *Kudryavcev-Platonov*. Another author, K. Istomin polemized with Solovyov's attempt of rationalization of the truth of faith (see А.А. Ермичев, *История русской философии в журнале "Вера и разум"*, "Вестник Русской Христианской Гуманитарной Академии" 2008, vol. 9(2), p. 150).

and practical needs. On the contrary, it aids them. "It heals any infirmity of reason caused by sin."[15]

Let us quote the opinion of M.A. Ostroumov, who wrote in his philosophy coursebook for students that "religious faith will show philosophy the ways of its studies, and philosophy will strengthen and clarify the faith, it will detach faith from misconceptions and superstitions."[16]

As it can be seen, philosophy was generally treated in the ecclesiastical-academic milieu as a field which was auxiliary or even ancillary to Revelation. It can be interpreted as a discipline searching for the understanding of the fundamentals of faith – in the spirit of Augustinian-Anselmian idea of *fides quaerens intellectum*. Russian philosophy eventually obtained autonomy in the works of the slavophiles and their opponents – occidentalists, especially among the thinkers of the Silver Age.[17] Nevertheless, the so-called academic philosophy bore great significance for the reception and the gradual development of the Russian tradition of philosophising that will retain the religious orientation present in the above-mentioned professors of ecclesiastical academies.

2. Scientific and Natural Apologetics

In the 19th century, Orthodox thinkers had to confront one more intensely developing field of study, namely the natural sciences. The biggest challenge of the time was biology,[18] and specifically the the-

---

[15] G. Florovsky, *Ways of Russian Theology*, trans. by R.L. *Nicholas*, <http://www.myriobiblos.gr/texts/english/florovsky_ways_chap5.html>.

[16] Cit. after: Л.Е. Шапошников, *Православие и философия: границы взаимодействия, op.cit.*, p. 70.

[17] Many of them, e.g. Vladimir Solovyov, and also Vasily Rozanov, Fr. Pavel Florensky or Fr. Sergei Bulgakov would later be accused – not without reason – of heterodoxy and sometimes even of heresy.

[18] Physical theories, such as Copernicanism, have never been subjected to condemnation in Eastern Orthodox Church, and what is more, this theory was taught even at the Kiev-Mohyla Academy. One of the authors writing about the teachings of Copernicus and Galileo with approval was, among others, Hryhorii Skovoroda (1722–1794).

ory of evolution. The figure considered to be the precursor of the reflection over the issue of the relations between science and faith is
Mikhail Lomonosov (1711–1765),[19] a scientist comprehensively educated in natural science and humanities, and the author of religious
poems. In the milieu of the ecclesiastical educational institutions –
academies and seminaries – science initially did not receive much attention; it was ignored, like philosophy used to be. The students of
the seminaries were only acquainted with practical information concerning agriculture that could be useful in the pastoral work of a village parish priest. Later, the departments of mathematics and natural sciences were created, but they did not exist for a long time. Due
to the reservations put forward by the over-procurator of the Holy
Synod, Count Dmitri Tolstoy (who insisted that in the academies only
theology should be taught), the natural sciences were removed from
the curriculum. The 1868 bill of the Ministry of Internal Affairs provided for teaching the physical and mathematical sciences only in the
Academy in Saint Petersburg, but this project was not accomplished.
In 1869 the authorities went as far as to close down the departments
of physics and mathematics in the academies. It was only due to the
efforts of the vice-chancellor of the Moscow Ecclesiastical Academy,
Aleksandr Gorsky,[20] supported by the Archbishop of Kamchatka In-
nokentii, the "substitute" department of scientific and natural apologetics was created on 12th October 1870. The former lecturers of
physics and cosmography acquainted students with the fundamental natural phenomena and scientific theories, but they did it only for
apologetic purposes.[21]

---

[19] See А.В. Солдатов, *Наука и религия в русской религиозной философии*, "Вестник Русской Христианской Гуманитарной Академии" 2007, vol. 8(2), pp. 142–143.
[20] Cf. Д.Ф. Голубинский, *Участие протоиерея А.В. Горского в деле учреждения
при Московской Духовной Академии кафедры естественно-научной апологетики*,
"Богословский вестник" 1900, vol. 3 (11), pp. 467–474.
[21] В. Заев, *Реформы духовных академий в XIX – начале XX в. I. Первая реформа духовных академий 1808–1814 гг.*, "Труди Київської Духовної Академії" 2008,
no. 9, pp. 350–352.

The awareness that one of the reasons for the conflict between faith and reason was the ignorance of the "overzealous" apologists of Christianity, who, "having the best intentions, usually possessed only a very superficial knowledge about science and nature"[22] enforced the revision of the curriculum in the ecclesiastical academies. At the same time, it was emphasized – like it was done as far back as by the early Christian writers, e.g. St. Augustine – that reason comes from God, and thus the attacks on knowledge (including science) in fact imply questioning the Divine intention. The professor of the Moscow Ecclesiastical Academy, Sergei Glagolev (1865–1937), called this tradition of combining the "classical rationality" with the Orthodox theology "the school of believing reason," making reference to the analogous expression of a slavophile Ivan Kireyevsky. The lecturers of the new discipline were supposed "not only to convince, but also to teach – obviously not the fundamentals of science in their entire range and content, since it is unfeasible with the use of the means of apologetics, but to *teach to believe* – first scientifically, then religiously."[23] In order to achieve it, they taught about the existence of God on the basis of His works (thus acquainting the students with teleological and cosmological arguments). At the same time, the professors were supposed to demonstrate the insufficiency of the "scientific faith," understood as the conviction about the rightness of the scientific doctrines, which – contrary to the invariable, irrefutable, or using modern language unfalsifiable dogmas of faith[24] – have barely the character of working hypotheses[25] and not the ultimate explanations of the Universe. Abandoning the strategy of isolation or the conflict between

---

[22] П.С. Страхов, *Богословие и естествознание (К вопросу о задачах естественно-научной апологетики)*, "Богословский вестник" 1908, vol. 1 (2), p. 258.

[23] *Ibid.*, p. 262.

[24] It must be emphasized that the Eastern Orthodox Church rejects the thesis about the evolution of dogmas. See e.g. Епископ Василий (Родзянко), *Теория распада вселенной и вера отцов. Каппадокийское богословие – ключ к апологетике нашего времени. Апологетика XXI века*, Паломник, Москва 1996, <http://bishop-basil.org/russian/works/book/part1.shtml>.

[25] The status of a hypothesis was ascribed, among others, to Darwin's theory on the origin of species developed at the time.

science and religion in favour of the co-operation between them was to facilitate – in the opinion of scientific and natural apologists – to realise the grand, noble purpose of comprehending God and the universe created by Him. At the same time, there was a call for creating "Christian" or, to be more specific, "Orthodox" philosophy which would raise the question about the relation of reason and faith in the spirit of the Eastern Orthodox Church.[26]

It ought to be emphasized that mathematics and natural sciences were taught not only in the academies, but also in the seminaries[27] (which after all provided secondary education), but only to a limited extent, on an incomparably lower level than in *gymnasia* preparing students for studies at the university. Professor V. Javorsky, in the journal "Богословский вестник", deplored the fact that the lecturers of the physical and mathematical sciences had at their disposal only 3 hours of algebra per week in the first year, 3 hours of geometry per week both in the second and the third year, and 3 hours of physics in the fourth year. After 1867, due to further reforms, natural history, trigonometry, and astronomy were removed from the curriculum, and the remaining subjects were limited to the minimum. Javorsky also complained about the lack of practice, the lack of the requirement for written assignments, and poor equipment in the laboratories, necessary for performing the experiments and exemplary lessons. As a result, a seminary graduate – the future village parish priest – had difficulty explaining even the most common physical phenomena which he encountered in his pastoral work, for instance the meteorological phenomena. And really,

> The alumnus of the seminary (…) is supposed to be a fighter in the world… In order to be worthy of his status, he ought to possess the sufficient level of intellectual development… As an apologist

---

[26] Cf. И.С. Вевюрко, *Научная рациональность и православное богословие в трудах мыслителей русских духовных школ начала XX века*, <www.bogoslov.ru/text/287359.html>.

[27] The equivalent of the lower seminaries in Western countries.

of faith, a priest must know natural sciences. A theologian often happens to touch the fundamentals of the natural sciences. Young theologians almost from the very beginning hear and know that the lack of faith and the negative tendencies of the contemporary world must be fought using their own tools... Therefore, the fighters must be given these tools...[28]

As it can be seen, some of the exponents of the Eastern Orthodox Church were conscious of the significance of the secular education for the clergy, although they first of all emphasized the apologetical purpose of studying mathematical and natural sciences, directed at the defence of the fundamentals of faith in the face of positivism, materialism and atheism. Academy professors also interpreted particular scientific theories in the spirit of harmony between faith and reason which they called for, even though that "concordance" often existed at the expense of not abiding by the competence of science, and subordinating its facts to the unbending dogmas of Christianity.

Let us remind ourselves of two standpoints of the search for the agreement between science and religion present in academic philosophy. The professor of philosophy, Victor Kudryavcev (1828–1891) wrote, that the settlement of the question about the origin of the Universe belongs to the natural sciences (astronomy, geology, paleontology and biology), adding that science does not entirely exhaust the subject, as it only explores the empirical world.[29] At the same time, Kudryavcev was inclined to acknowledge Darwin's theory as a plausible explanation of the origins of human species. Another lecturer of scientific and natural apologetics, Dimitri Golubinsky (1832–1903, the son of an exponent of academic philosophy, Fyodor Golubinsky)

---

[28] В. Яворский, *Кафедра "физико-математических наук" в духовных семинариях (Несколько слов и мыслей по поводу ожидаемой реформы духовно-учебных заведений)*, "Богословский вестник" 1902, vol. 2 (7/8), pp. 573–574.

[29] Cf. В.Д. Кудрявцев, *Регрессивная и прогрессивная теория происхождения мира*, "Богословский вестник" 1892, vol. 1 (1), pp. 19–20.

taught that "science ought to be conscious – personified by its lecturers – of its helplessness about certain issues," at the same time adding a controversial thesis that "numerous phenomena of visible nature cannot be ultimately explained barely in terms of nature, just the opposite, one should acknowledge the supernatural action of the almighty Creator."[30] The statement about the limitation of science, as well as emphasizing that Divine action in the world does not evoke controversy, however, the methodologically misformulated assertion about the impossibility of scientific explanation of the empirical world within the world itself, immediately violates the independence of science and reduces it to the role of an auxiliary discipline, subordinate to theology.

Both theologians – quite properly – indicated the insufficiency of scientific explanation, but they differed in the evaluation of scientific facts, and specifically Darwin's theory. Contrary to V. Kudryavcev, who recognized the cognitive significance of science, D. Golubinsky seemed to be satisfied with the claim that "science does not contradict religion," thus employing, so to say, the principle of decontradictification, and in the conflictual situation he decidedly rejected the theories which seemed to undermine the fundamentals of faith. Due to this, he described "the views of the Darwinists" as "nonsensical," "unreasonable," "unproved" and "unsupported," thus violating the autonomy and the cognitive value of science. As he wrote:

> To certain detailed questions concerning the creation of the world, it
> is safer to reply in the following manner:
>> This we do not know.[31]

Apophatism – not only religious, but also scientific – was, for Golubinsky and many other scientific and natural apologists, the best and

---

[30] Д.Ф. Голубинский, *Открытое письмо к N.N. по поводу вопросов о сотворении мира*, "Богословский вестник" 1895, vol. 3 (8), pp. 202–203.
[31] *Ibid.*, p. 207.

the most reliable strategy in debatable issues. The mysteriousness of the "exceptionally intricate" subject matter which the work of creating the world is, hindered the progress towards making attempts to confront the positive scientific discoveries.

Some authors, for instance Sergey Glagolev sought the purpose of scientific and natural apologetics in the scientific justification of the fundamentals of faith.[32] It is obvious that it is an undertaking which is doomed to failure, since it violates the fields of interest of science and religion. The exponents of the new discipline not always guided themselves with appropriate methodology in their experiments. As a result, apologetics practised in such manner disregarded scientific facts and set science in the background, and that is why the assurances about initiating the dialogue between science and religion remained within the sphere of wishful thinking.

3. The Issue of the Relationship between Science and Religion in the 20th Century

After the 1917 revolution, the question of the relationship of science and faith became an especially urgent problem in the USSR, since the communist propaganda relied exactly on the scientific facts. This is why numerous clergymen of the Eastern Orthodox Church reattempted the apology of faith from the accusation of inconsistency with the discoveries of the positive science. The problem in question was the subject of deliberation of Nikolai Fyoletov (1891–1943), the author of posthumously published *Outline of Christian apologetics*, in which the theologian investigated such issues as the origin of the Universe and the man, the problem of miracles, natural laws etc. from

---

[32] See О. Мумриков, *Естественно-научная апологетика как целостная дисциплина: общий обзор*, "Вестник Православного Свято-Тихоновского богословского института. IV: Педагогика. Психология" 2009, vol. 4 (15), p. 28.

Christian perspective, and also of Father Luka (Voyno-Yasenetsky, 1877–1961, the author of the works *Science and religion* and *Spirit, Soul and body*).

The works which deserve special attention are the publications of the Russian emigrant theologians working in Paris: Vassily Zienkovsky (1881–1962), Vladimir Lossky (1903–1958) and Georges Florovsky (1893–1979). Zienkovsky devotes the first part of his *Apologetics* to the relations of the Christian faith and the contemporary scientific knowledge. The author emphasizes that the conflict between faith and reason occurs only in the situation of the isolation of the latter from the tradition of the Church. It is not clear whether it means that scientific cognition ought to be – according to Zienkovsky – subordinated to the doctrine of the Church. For Zienkovsky, on one hand, writes about the autonomy of science:

> Eastern Orthodox faith creates wide space for *exploring* nature. (…)
> The freedom of research is the essential condition of scientific work.[33]

On the other hand, the author expresses the opinion about the superiority of theological cognition over the scientific one: "We are conscious of our duty and right of exploring and explaining the natural phenomenon in the light of Christ."[34] Why? Out of the simple reason that Zienkovsky taught that science explores the results of Divine action, and thus indirectly leads to superior religious cognition.

> The problem of God's presence in the world concerns the sphere to which both theology and science aspire, since it is about God's action *in the world* that reveals itself as much by religious contemplation as through scientific research. (…) Indeed, science explores nature as if God's participation in the life of the world never and nowhere

---

[33] В.В. Зеньковский, *Основы христианской философии* [*Basics of Christian Philosophy*], vol. 1, Канон, Москва 1997, pp. 88–89.
[34] *Ibid.*, p. 101.

became apparent. However, while science does not sense the perplexity of such attitude, for Christian theology it is obviously a dead end. (…) Exploring nature must essentially lead to the *metaphysics* of the world (…).[35]

According to Zienkovsky, the search for the relationship between science and religion, the exploration of the empirical and the extra--empirical is characteristic of Russian thought in a unique way.[36]

As far as the conflict between science and religion is concerned, Zienkovsky admits priority and rightness to the fundamentals of faith, writing about the hypothetical character of science. And it is this hypotheticality which hinders attempts at an explicit, definite and complete settlement of the fundamentals of faith and the scientific facts:

If some statements of the contemporary knowledge can in no way correspond with the Christian doctrine of faith, there is nothing tragic for either side. Scientific ideas and generalisations are continuously *im Werden*, certain hypotheses are replaced by others, certain ideas give way to another ones.[37]

Thus, Zienkovsky shuns the methodologically erroneous position of concordism which was characteristic to numerous exponents of Rus-

---

[35] В.В. Зеньковский, *Об участии Бога в жизни мира*, [in:] *idem, Собрание сочинений*, vol. 2: *О православии и религиозной культуре*, Русский путь, Москва 2008, pp. 345, 350, 356. Cf. *idem, Основы христианской философии*, vol. 1, p. 64: "Whatever we would discover in the world, we discover owing to the Divine presence in the world. (…) Any cognition 'the relation to the Absolute'."

[36] Cf. В.В. Зеньковский, *О мнимом материализме русской науки и философии*, [in:] *idem, Собрание сочинений*, vol. 1: *О русской философии и литературе*, Русский путь, Москва 2008, pp. 316–317.

[37] В.В. Зеньковский, *Основы христианской философии, op.cit.*, vol. 1, p. 89. Cf. *idem, Апологетика* [*Apologetics*], <http://www.klikovo.ru/db/book/msg/4132>: "Science, in its progress, must either replace some hypotheses by another ones, or modify them to such extent that in fact they become new ones. (…) However, the text of the Bible remains unchanged."

sian thought at the turn of the 20th century, although he does not avoid the temptation of subordinating science to religion.

As opposed to Zienkovsky, Vladimir Lossky was not occupied with the issue of the relationship between science and religion as such. The question of scientific cognition appears marginally in his works, in connection with the subject of apophatism, the key issue for the Eastern Orthodox Church. This Orthodox theologian elaborated on the motive (already present in the academic philosophy) of the limitation of human cognition (including the scientific cognition concerning the exploration of nature) on the basis of the texts by the Church Fathers. He wrote:

> For St. Basil, not the divine essence alone but also created essences could not be expressed in concepts. (…) There will always remain an "irrational residue" which escapes analysis and which cannot be expressed in concepts; it is the unknowable depth of things, that which constitutes their true, indefinable essence.[38]

Lossky – like other Paris theologians such as Vassily Zienkovsky or Georges Florovsky[39] – by no means rejected the possibility of getting to know the world and God, but he emphasized that it concerns only Divine actions – powers, energies and not His essence. Those energies are present in the world, thus exploring the mysteries of nature is an indirect way to knowing the Creator.

At present (after the Church has left the underground) the issue of the relationship between faith and science is one of the most widely discussed, both in the Russian academic milieu (the theological, philosophical and scientific ones), and in the press and other mass media.

---

[38] V. Lossky, *The Mystical Theology of the Eastern Church*, transl. by members of the Fellowship of St. Alban and St. Sergius, St Vladimir's Seminary Press, Crestwood – New York 1976, p. 33. Cf. К.В. Преображенская, *Богословие и мистика в творчестве Владимира Лосского*, Издательство СПбГУ, Санкт-Петербург 2008, pp. 34–35.

[39] Cf. G. Florovsky, *The Idea of Creation in Christian Philosophy*, "Eastern Churches Quarterly" 1949, vol. 8 (2), pp. 53–77.

Numerous conferences, panel discussions, and debates on this subject have been organised in various contexts, among others the methodological, biblical or philosophical ones.[40] Also the hierarchs of the Eastern Orthodox Church take part in the discussion over this issue.[41] It is noteworthy that some scholars, in defending the faith (sometimes from the alleged menace constituted by science), quote – in accordance with the rule *consensus patrum* ("the consensus of the Fathers"), still valid in the Eastern Orthodox Church – particular opinions of the early Christian authors on for example the origin of man (usually in the spirit of creationism). Other, more open and discerning thinkers teach that "the consensus of the Fathers" should be looked for "not in the exterior phrases but in *what concerns the spirit* – the appropriate attitude to the interpretation of certain passages of the Holy Scripture",[42] rejecting the literal exegesis of the Bible and studying natural sciences. According to the author of the above citation "some of the contemporary apologists (…) attacking the 'secular science' instead of utilizing it for the benefit of the Church where it is useful, act in an extremely unreasonable manner."[43]

The standpoints concerning the relationship between science and faith are exceptionally varied – from the extreme concordism on one hand to the extreme separatism on the other, through numerous more or less successful attempts of a dialogue or of subordinating

---

[40] I am going to mention two well-known contemporary textbooks of scientific and natural apologetics: Е. Порфирьев, *Православная естественно-научная апологетика*, Краснодар 2006; А.И. Осипов, *Путь Разума в поисках истины* (several editions), Москва (both the authors defend the position of creationism); В.Д. Захаров, Естественно-научная апологетика, [in]: Ю.С. Владимиров, А.В. Московский (eds.), *Христианство и наука. Сборник докладов конференции*, Московски Патриархат, Москва 2001, pp. 197–225.

[41] See e.g. Metropolitan Filaret of Minsk and Slutsk, *God and Physical Cosmology*, "Faith and Philosophy. Journal of the Society of Christian Philosophers" 2005, vol. 22, no. 5, pp. 521–527.

[42] О. Мумриков, *Церковь и естественнонаучные картины мира: проблемы рецепции*, <http://www.mpda.ru/site_pub/129001.html>. See also Special Issue of the journal "Vstrecha" (*Встреча*), vol. 3 (21) 2005, edited by the students of the Moscow Theological Academy.

[43] О. Мумриков, *Церковь и естественнонаучные картины мира…*, *op.cit.*

the scientific cognition to the religious one. Nevertheless, the sole fact of the broad interest in the problem discussed here allows one to cherish the hope that a comprehensive and impartial quest for the answers to the significant questions vexing our contemporaries will also be continued.